Harmonic Quasiconformal Mappings and Hyperbolic Type Metrics

Vesna Todorčević

Harmonic Quasiconformal Mappings and Hyperbolic Type Metrics

Vesna Todorčević
Faculty of Organizational Sciences
University of Belgrade
Belgrade, Serbia

Mathematical Institute
Serbian Academy of Sciences and Arts
Belgrade, Serbia

ISBN 978-3-030-22593-3   ISBN 978-3-030-22591-9   (eBook)
https://doi.org/10.1007/978-3-030-22591-9

Mathematics Subject Classification: 30C65, 30C62, 31B05, 31B15, 31B25

This Springer imprint is published by the registered company Springer Nature Switzerland AG.
The registered company address is: Gewerbestrasse 11, 6330 Cham, Switzerland

Preface

The goal of this book is to present a research area in Geometric Function Theory concerned with harmonic quasiconformal mappings and hyperbolic type metrics defined on planar and multidimensional domains. The classes of quasiconformal and quasiregular mappings are well-established areas of study in this field as these classes are natural and fruitful generalizations of the class of analytic functions in the planar case. Harmonic mappings are another natural generalization of conformal mappings and analytic functions and form another well-established class. Injective quasiregular mappings are quasiconformal, and conformal mappings are both harmonic and quasiconformal. On the other hand, harmonic mappings are smooth, and when quasiregular, they are also locally quasiconformal independently of the dimension. So in higher dimensions, the study of the class of mappings that are both harmonic and quasiconformal suggests itself. It turns out that while this seems at first a rather restrictive class, the study of this class uncovers new and unexpected phenomena and is today recognized as an important research area in Geometric Function Theory. The book contains many concrete examples, as well as detailed proofs and explanations of motivations behind given results, gradually bringing the reader to the forefront of current research in the area. The book is written for a wide readership from graduate students of mathematical analysis to researchers working in this or related areas who want to learn the tools or work on the open problems, many of which are listed in various parts of the book, especially in the last chapter. An extensive bibliography of the field is also given for the readers who wish to explore deeper into the results presented in the book or related results that are not covered here. Prerequisite knowledge for reading this book includes the basic knowledge of real and complex analysis, harmonic functions, and the topology of metric spaces. The book is primarily based on research done in the last 12 years, starting with the author's master and doctoral dissertations and followed by a number of papers that are either single authored or jointly authored with other experts in the field. The author is therefore grateful to all her collaborators and other mathematicians who have built this research area and have shared their expertise with enthusiasm. Without their help, this book would not have come into its existence.

Belgrade, Serbia   Vesna Todorčević
January 2019

Contents

Notation

$\mathbb{Z}$	Set of integers
$\mathbb{N}$	Set of positive integers
$\mathbb{R}$	Set of real numbers
$\mathbb{C}$	Set of complex numbers
$|z|$	Modulus of the complex number z
$\arg(z)$	Argument of the complex number z
$\mathbb{R}^n$	n-dimensional Euclidean space
$\overline{\mathbb{R}^n} = \mathbb{R}^n \cup \{\infty\}$	Möbius space
$|x|$	Euclidean norm of a vector $x \in \mathbb{R}^n$
$\mathbb{B}^n(x, r)$	Open ball centered at $x \in \mathbb{R}^n$ with radius $r > 0$
V_n	Volume of the n-dimensional unit ball
$S^{n-1}(x, r)$	Sphere centered at $x \in \mathbb{R}^n$ with radius $r > 0$
ω_{n-1}	$(n-1)$-dimensional measure of $S^{n-1}(0, 1)$
$\mathbb{H}^n$	Poincare half-space
S_ρ	Planar angular domain
$P(a, t)$	$(n-1)$-dimensional hyperplane
$\mathscr{GM}(\overline{\mathbb{R}^n})$	Group of Möbius transformations
$M_3(\mathbb{R})$	Set of square matrices of order 3
$\pi(x)$	Stereographic projection
$q(x, y)$	Spherical (chordal) distance between x and y
$Q(x, r)$	Spherical ball
$|a, b, c, d|$	(Absolute) cross ratio
a^*	Image of the point a under an inversion on S^{n-1}
$D(a, M)$	Hyperbolic ball with center a and radius M
$\mathrm{dist}(x, A)$	Distance of a point $x \in \mathbb{R}^n$ to a set $A \subseteq \mathbb{R}^n$
∂A	Boundary of a set $A \subseteq \mathbb{R}^n$
$diam(A)$	Diameter of a set $A \subseteq \mathbb{R}^n$
χ_A	Characteristic function of a set $A \subseteq \mathbb{R}^n$
$l(\gamma)$	Length of the curve γ
$\rho(x, y)$	Hyperbolic distance between x and y
j_D	Distance ratio metric

k_D	Quasihyperbolic metric
δ_G	Seittenranta metric
α_G	Apollonian metric
s_G	Triangular ratio metric
λ_G	Ferrand metric
μ_G	Modulus metric
$\Delta(E, F; G)$	Family of all closed nonconstant curves joining E, F in G
$R_{G,n}(s)$	Grötzch ring
$R_{T,n}(s)$	Teichmüller ring
$\gamma(s), \gamma_n(s)$	Capacity of $R_{G,n}(s)$
$\tau(s), \tau_n(s)$	Capacity of $R_{T,n}(s)$
$\varphi_K(r), \varphi_{K,n}(r)$	Special function related to the Schwarz lemma
$\log \Phi_n(s)$	Modulus of the Grötzsch ring
$\log \psi_n(s)$	Modulus of the Teichmüller ring
λ_n	Grötzsch ring constant
$mod(R)$	Modulus of a ring
$cap(R)$	Capacity of a ring
$\mathrm{cap}_W(K)$	Wiener capacity
$p\text{-}cap\, E, cap\, E$	(p-)capacity of a condenser
$\Lambda_\alpha(F)$	α-dimensional Hausdorff measure of F
$N(f, A, y)$	Number of preimages of point y in A under f
$Lip(f)$	Lipschitz constant of f
$K(f), K_O(f), K_I(f)$	Maximal, outer, and inner dilatation of f
$H(x, f)$	Linear dilatation of a mapping f at x
$M_p(\Gamma), M(\Gamma)$	(p-)modulus of the curve Γ
$Id_K(\partial \mathbb{B}n)$	K-qc maps with identity boundary values
$C^1(C^2)$	Class of functions with continuous first-order (second-order) derivatives
$C_C^\infty(\Omega)$	Class of compactly supported functions with derivatives of all orders
∇f	Gradient of mapping $f : \Omega \longrightarrow \mathbb{R}^n$
Df	Weak derivative of mapping f
$J_f(x)$	Jacobian of mapping $f : \Omega \longrightarrow \mathbb{R}^n$ at x
$\alpha_f(z)$	Average derivative
$\mathcal{H}_u$	Hessian of u
$\|Du(x)\|$	Hilbert–Schmidt norm
Δf	Laplacian of f
$L^p(\Omega)$	Lebesgue space
$\|f\|_{L^p}$	L^p-norm of function f
L_{loc}^p	Local Lebesgue space
$W^{1,p}(\Omega)$	Sobolev space
$W_{loc}^{1,p}(\Omega)$	Local Sobolev space
$P(x, \xi)$	Poisson kernel for the unit ball
$G_{\mathbb{B}^n}(x, y)$	Green function on the unit ball

$\mathscr{I}_s h$	Riesz potential of order s
$QNS_K(\Omega)$	Class of all quasi-nearly subharmonic functions for a fixed K
$QNS(\Omega)$	Class of all quasi-nearly subharmonic function defined in Ω
$RO(\Omega)$	Class of regularly oscillating functions in Ω
$\|u\|_{QNS}$	QNS-norm of function u
$\|u\|_{RO}$	RO-norm of function u
$\|u\|_{BMO}$	BMO-norm of function u

Preliminaries

The background material necessary for reading this book can be found in standard texts of this area of mathematics such as [23, 138] and [155]. We shall also try to follow the standard terminology and notation as much as possible, but in this chapter, we list the notation and the terminology that are specific for this book.

A pair (X, d) is called a metric space if $X \neq \emptyset$ and $d : X \times X \rightarrow [0, +\infty)$ satisfies the following four conditions:

(M1) $d(x, y) \geqslant 0$, for all $x, y \in X$,
(M2) $d(x, y) = 0$ iff $x = y$,
(M3) $d(x, y) = d(y, x)$ for all $x, y \in X$,
(M4) $d(x, z) \leqslant d(x, y) + d(y, z)$ for all $x, y, z \in X$.

A pair (X, d) is called a pseudometric space if $X \neq \emptyset$ and $d : X \times X \rightarrow [0, +\infty)$ satisfies the following four conditions:

(M1) $d(x, y) \geqslant 0$, for all $x, y \in X$,
(M2') $d(x, x) = 0$,
(M3) $d(x, y) = d(y, x)$ for all $x, y \in X$,
(M4) $d(x, z) \leqslant d(x, y) + d(y, z)$ for all $x, y, z \in X$.

The inner product $\langle (a_1, \ldots, a_n), (b_1, \ldots, b_n) \rangle$ is defined to be equal to the sum $a_1 b_1 + \cdots + a_n b_n$, and we let $|a| = \sqrt{\langle a, a \rangle}$.

Let (X, d_1) and (Y, d_2) be metric spaces, and let $f : X \rightarrow Y$ be a continuous mapping. Then, we say that f is uniformly continuous if there exists an increasing continuous function $\omega : [0, \infty) \rightarrow [0, \infty)$ with $\omega(0) = 0$ such that

$$d_2(f(x), f(y)) \leq \omega(d_1(x, y)) \text{ for all } x, y \in X.$$

We call the function ω the modulus of continuity of f. If there exist $C, \alpha > 0$ such that $\omega(t) \leq C t^\alpha$ for all $t \in (0, 1)$, we say that f is Hölder continuous with

Hölder exponent α. If $\alpha = 1$, we say that f is Lipschitz with the Lipschitz constant C or simply C-Lipschitz. If f is a homeomorphism and both f and f^{-1} are C-Lipschitz, then f is C-bi-Lipschitz or C-quasiisometry, and if $C = 1$, we say that f is an isometry. These conditions are said to hold locally if they hold for each compact subset of X.

A very special case of these mappings are the isometries. Recall that if (X_1, d_1) and (X_2, d_2) are metric spaces and $f : X_1 \to X_2$ a homeomorphism, then we call f an *isometry* if $d_2(f(x), f(y)) = d_1(x, y)$ for all $x, y \in X_1$.

We use the notation

$$B^n(x, r) = \{y \in \mathbb{R}^n : |x - y| < r\},$$

$$S^{n-1}(x, r) = \{y \in \mathbb{R}^n : |x - y| = r\},$$

$$\mathbb{H}^n = \{(x_1, \ldots, x_n) \in \mathbb{R}^n : x_n > 0\}$$

and abbreviations $B^n(r) = B^n(0, r)$, $\mathbb{B}^n = B^n(1)$, $S^{n-1}(r) = S^{n-1}(0, r)$, and $S^{n-1} = S^{n-1}(1)$.

Recall that the volume of the n-dimensional unit ball can be expressed as

$$V_n = \frac{\pi^{\frac{n}{2}}}{\Gamma(\frac{n}{2} + 1)},$$

where Γ is Euler gamma function. Recall also that if by ω_{n-1}, we denote the $(n-1)$-dimensional measure of the sphere S^{n-1}, then

$$\omega_{n-1} = nV_n.$$

A Möbius transformation in complex plane is a mapping of the form

$$z \mapsto \frac{az + b}{cz + d},$$

where $a, b, c,$ and d are complex constants such that $ad \neq bc$.

The homeomorphism $f : D \longrightarrow D'$ is called conformal, where D and D' are domains in $\mathbb{R}^n$, if f is $C^1(D)$, if $J_f(x) \neq 0$ for all $x \in D$, and if $|f'(x)h| = |f'(x)| \, |h|$ for all $x \in D$ and $h \in \mathbb{R}^n$. An anticonformal is the complex conjugate of a conformal mapping.

If D and D' are domains in $\overline{\mathbb{R}^n}$, a homeomorphism $f : D \longrightarrow D'$ is conformal if its restriction to $D \setminus \{\infty, f^{-1}(\infty)\}$ is conformal. This definition of conformal homeomorphism is preferable to others, since if f is conformal, then f^{-1} is conformal as well because $J_f(x) \neq 0$.

Let D and D' be domains in $\overline{\mathbb{R}}^n$. We call a C^1-homeomorphism $f : D \longrightarrow D'$ *sense-preserving* (orientation-preserving) if $J_f(x) > 0$ for all $x \in D \setminus$

$\{\infty, f^{-1}(\infty)\}$. If $J_f(x) < 0$ for all $x \in D \setminus \{\infty, f^{-1}(\infty)\}$, then we call f *sense-reversing* (orientation-reversing).

The mapping $f : D \longrightarrow D'$ is regular at $x \in D$ when f is differentiable at x and $J_f(x) \neq 0$.

A bijective mapping $f : G_1 \rightarrow G_2$ between domains G_1 and G_2 in $\mathbb{R}^n$ is a diffeomorphism if f and f^{-1} are continuously differentiable. Note that if f is a diffeomorphism, then $J_f(x) \neq 0$ for all $x \in G_1$.

Let D and D' be domains in $\mathbb{R}^n$. The Hessian of a function $f : D \longrightarrow \mathbb{R}^n$ is defined to be

$$\det\left(\left[\frac{\partial^2 f}{\partial x_i \partial x_j}\right]_{i,j=1}^n\right).$$

A function $f : D \longrightarrow \mathbb{R}$ is called harmonic if $f \in C^2(D)$ and

$$\Delta f := \sum_{i=1}^n \frac{\partial^2 f}{\partial x_i^2} = 0.$$

The operator Δ is called Laplacian. The set of harmonic functions forms a linear space.

A complex valued function $f : \Omega \longrightarrow \mathbb{C}$, where Ω is domain in $\mathbb{R}^n$, is called harmonic if both of its real and imaginary parts are harmonic functions in Ω.

Let X, Y be domains in $\mathbb{R}^n$. If a function $f : X \longrightarrow Y$ belongs to $C^2(X)$ and if each of its coordinate functions is harmonic, then we say that f is harmonic.

Note that the real and imaginary parts of the complex analytic function are harmonic functions.

Note that if f is harmonic and g analytic, then the composition $f \circ g$ is a harmonic function, but the composition $g \circ f$ need not be harmonic in general. Note also that all conformal mappings are analytic and that, in the plane, all analytic functions are harmonic. Additionally, in the plane, each harmonic function is locally the real (imaginary) part of some analytic function.

The Harnack inequality for positive harmonic function $u : G \longrightarrow (0, \infty)$ states that for every $s \in (0, 1)$, there is $C \geq 1$ such that

$$\max_{z \in B_x} u(z) \leq C \min_{z \in B_x} u(z),$$

holds, whenever $B^n(x, r) \subseteq G$ and $B_x = \overline{B}^n(x, sr)$.

Green's function on the unit ball is defined by

$$G(x, y) = \Phi(y - x) - \Phi(|x|(y - x^*)), \quad x, y \in \mathbb{B}^n, \quad x \neq y.$$

Here, ω_{n-1} is the $(n-1)$-dimensional measure of S^{n-1}, $x^* = |x|^{-2}x$, and the function

$$\Phi(x) = \begin{cases} -\frac{1}{2\pi} \log |x|, & n = 2, \\ \frac{1}{n(n-2)\omega_{n-1}} \frac{1}{|x|^{n-2}}, & n \geq 3, \end{cases}$$

defined for $x \in \mathbb{R}^n$, $x \neq 0$ is the fundamental solution of the Laplace's equation.

Recall that

$$P(x, \xi) = \frac{1 - |x|^2}{|x - \xi|^n}$$

is the Poisson kernel for the unit ball in $\mathbb{R}^n$.

Recall also that a continuous function $u : G \longrightarrow \mathbb{R}$ defined on a domain $G \subset \mathbb{C}$ is subharmonic if for all $z_0 \in G$, there exists $\varepsilon > 0$ such that

$$u(z_0) \leq \frac{1}{2\pi} \int_0^{2\pi} u(z_0 + re^{it}) \, dt \text{ for } 0 < r < \varepsilon. \tag{$*$}$$

A continuous function $u : G \longrightarrow \mathbb{R}$ defined on a domain $G \subset \mathbb{C}$ is superharmonic if $-u$ is subharmonic.

Let a and b are reals such that $a < b$. For path $\alpha : [a, b] \longrightarrow \mathbb{R}^n$, we define its length $l(\alpha)$ as the supremum of values of the form

$$\sum_{i=1}^{n} |\alpha(t_i) - \alpha(t_{i-1})|$$

where $n \in \mathbb{N}$ and

$$a = t_0 < \cdots < t_n = b.$$

If $l(\alpha) < \infty$, then we say that α is a rectifiable path.

If α is a rectifiable path, then there is a unique path $\alpha^0 : [0, l(\alpha)] \longrightarrow \mathbb{R}^n$ such that there is a continuous increasing function $h : [a, b] \longrightarrow [0, l(\alpha)]$ such that $\alpha = \alpha^0 \circ h$ and for any $t \in [0, l(\alpha)]$, $l(\alpha^0|_{[0,t]}) = t$. We call the path α^0 a normal representation of α.

Let $I \subseteq \mathbb{R}$ is interval and $f : I \longrightarrow \mathbb{R}^n$. We say that f is absolutely continuous if for every $\varepsilon > 0$, there is $\delta > 0$ such that for every $m \in \mathbb{N}$ and for every sequence $(a_1, b_1), \ldots, (a_m, b_m)$ of pairwise disjoint subintervals of I, we have that

$$\sum_{i=1}^{n} |a_i - b_i| < \delta \Rightarrow \sum_{i=1}^{n} |f(a_i) - f(b_i)| < \varepsilon.$$

A Jordan curve or simple closed path in metric space (X, d) is a continuous (with respect to the metric d) mapping $\gamma : [0, 1] \longrightarrow X$ such that for all $x, y \in [0, 1]$ with $x < y$,

$$\gamma(x) = \gamma(y) \Leftrightarrow x = 0 \wedge y = 1.$$

A plane domain is Jordan if its boundary is a Jordan curve. For higher dimensions n, we say that a domain in $\mathbb{R}^n$ is a Jordan domain if its boundary is homeomorphic to the unit sphere.

We denote the α-dimensional Hausdorff measure of a set $F \subset \mathbb{R}^n$ by $\Lambda_\alpha(F)$. Recall that

$$\Lambda_\alpha^\delta(F) = \inf\{\sum_{i=1}^{\infty} d(U_i)^\alpha\},$$

where the infimum is taken over all countable coverings of F by sets U_i with $d(U_i) < \delta$, then set $\Lambda_\alpha(F) = \lim_{\delta \to 0} \Lambda_\alpha^\delta(F)$. The Hausdorff dimension of a set F is defined as follows

$$\dim_H(F) = \inf\{\alpha : \Lambda_\alpha(F) < \infty\}.$$

The β-dimensional Hausdorff content is defined to be

$$\Lambda^\beta(E) = \inf\left\{\sum_{i=1}^{\infty} r_i^\beta\right\},$$

where the infimum is taken over all coverings of $E \subseteq \mathbb{R}^n$ with countably many (Euclidean) balls of radii r_i.

Let $G \subset \mathbb{R}^n$ be a domain, and let $w : G \longrightarrow (0, \infty)$ be continuous. For given $x, y \in G$, let

$$d_w(x, y) = \inf\{l_w(\gamma) : \gamma \in \Gamma_{xy}, l(\gamma) < \infty\}, \quad l_w(\gamma) = \int_\gamma w(\gamma(z))|dz|.$$

It turns out that this is a metric on G. If a length-minimizing curve exists, it is called a geodesic.

For a proper subdomain G of $\mathbb{R}^n$, the quasihyperbolic length of a rectifiable curve γ in G is given by

$$l_k(\gamma) = \int_\gamma \frac{|dz|}{d(z, \partial G)}.$$

The quasihyperbolic distance between points x and y from G is the infimum of quasihyperbolic lengths over all rectifiable curves in G joining x and y.

For an easy reference, we also record the Bernoulli inequality,

$$\log(1 + as) \leqslant a \log(1 + s), \quad a \geqslant 1, \; s > 0.$$

Chapter 1
Introduction

Geometric Function Theory began as a branch of Complex Analysis dealing with geometric aspects of analytic functions, but has since grown considerably, both in scope and in methodology. It considers, for example, the class of quasiregular mappings proven to be a natural and especially fruitful generalization of analytic functions in the planar case. Another class considered is the class of quasiconformal mappings characterized by the property[1] that there is a constant $C \geq 1$ such that infinitesimal spheres are mapped onto infinitesimal ellipsoids in such a manner that the ratio of the longest axis to the shortest axis is bounded from above by C. Injective quasiregular mappings are quasiconformal and conformal mappings in the plane are both harmonic and quasiconformal. Moreover, harmonic mappings are smooth and if they are also quasiregular they are locally quasiconformal in higher dimensions. This gives us a motivation to study harmonic quasiconformal mappings in higher dimensions. Today the study of these classes of mappings is recognized as an important research area of Geometric Function Theory.

In more precise analytic terms, a quasiconformal map $f : X \rightarrow Y$ is a homeomorphism of two domains in $\mathbb{R}^n$ that is differentiable almost everywhere, such that f belongs to Sobolev space $W^n_{1,loc}$ and there is a uniform bound K on the ratio of the largest and smallest absolute value of eigenvalue of a differential of f, valid almost everywhere in X. There are many other alternative definitions of quasiconformal mappings that use, for example, moduli of families of curves or linear dilatation which are more geometric in nature showing that quasiconformality is a fruitful notion. The equivalence of geometric and analytic definitions of quasiconformal mappings has been established for quite some time. The two-dimensional quasiconformal theory was developed by mathematicians including Lars Ahlfors, Lipman Bers, Oswald Teichmüller, Frederick Gehring, and William Thurston in the 20th century. Quasiconformal mappings have compactness properties similar to conformal mappings. For instance, they can be used to form

---

[1]In Chap. 2, we will formally define the notions of quasiconformal and quasiregular mappings.

© Springer Nature Switzerland AG 2019
V. Todorčević, *Harmonic Quasiconformal Mappings and Hyperbolic Type Metrics*,
https://doi.org/10.1007/978-3-030-22591-9_1

normal families of functions under quite general conditions, which gives them a special place in Geometric Function Theory. The theory of two-dimensional quasiconformal mappings has applications in numerous areas of mathematics such as Teichmüller theory of Riemann surfaces, Complex Dynamics [30] (the famous No Wandering Domains Theorem of Dennis Sullivan), low dimensional Topology, as well as in Physics (String Theory). Higher dimensional applications have been less versatile, due to the rigidity of quasiconformal mappings, discovered by George Mostow in 1968, [119].

Harmonic mappings are another natural generalization of conformal mappings and analytic functions. These are used extensively in the study of Teichmüller space. The first significant application was given by M. Wolf [161]. Later R. Schoen [139] posed a well-known conjecture stating that every quasisymmetric homeomorphism $u : \partial \mathbb{H}^2 \to \partial \mathbb{H}^2$ admits a unique harmonic quasiconformal extension $f : \mathbb{H}^2 \to \mathbb{H}^2$. This conjecture was resolved recently by V. Marković in [108] and his result can be used to find a canonical HQC representative for each class in the Universal Teichmüller space.

Since the theories of harmonic mappings and quasiconformal mappings are both well developed, it is of interest to consider how the corresponding results can be strengthened in the presence of both harmonicity and quasiconformality [1, 74, 75, 85, 102, 117]. While these definitions impose strong limitations, some of the results are unexpected and elegant. One such result is, for example, the preservation of boundary modulus of continuity in the case of the unit ball given in [15]. Harmonic quasiconformal (abbreviated as HQC) mappings in the plane were first introduced by Olli Martio in [110]. Today they are investigated both in the planar and in the multidimensional setting from several different points of view. Unfortunately, the powerful machinery developed for the plane is not available in the space, and so our approach is to combine the analytic and geometric aspects of the theory of quasiconformal mappings together with a number of tools from Harmonic Analysis. Among the topics considered in this book are the boundary behavior, including Hölder and Lipschitz continuity, and the more general moduli of continuity, behavior with respect to natural metrics, especially quasihyperbolic metric, distortion estimates, bi-Lipschitz properties with respect to different metrics, and characterization of boundary mappings. We shall also explain an array of tools used in this study such as the conformal moduli of families of curves, Poisson kernels, estimates from the theory of second order elliptic operators, notions of capacity, subharmonic functions, and Riesz potentials.

In Chap. 2 we introduce basic notions and examples on which the rest of the book relies. Conformal invariants, such as harmonic measure, hyperbolic distance, condenser capacity, and modulus of a curve family are fundamental tools of Geometric Function Theory. The theory of quasiconformal mappings in $\mathbb{R}^n, n \geq 2$ makes effective use of all these tools. For example, even the definition of quasiconformal mappings can be expressed in terms of moduli of families of curves. (For the planar case $n = 2$ the reader can see this in the L. Ahlfors book [5] and in the higher dimensional case in the J. Väisälä book [155].) The relationship between moduli of continuity of a harmonic quasiregular mapping on the boundary and inside the ball in dimension $n = 2$ is given in [88] and in higher dimensions in [15]. Chapter 2 also contains an example of a non-Lipschitz harmonic quasiconformal mapping on the unit ball. In [16] it is shown that for a wide range of domains, including those with uniformly perfect boundary, Hölder continuity on the boundary implies Hölder continuity (with the same exponent) inside the domain for the class of HQC mappings, a result which does not hold for the class of qc mappings.

In Chap. 3 we introduce some hyperbolic type metrics and explain their connection with the theory of qc mappings. This turns out to be a fruitful theme explored by a number of authors. The geometric properties of domains have been another central theme of research, especially in relation to the boundary phenomena. The geometric nature of the boundary is reflected by the quasihyperbolic metric introduced by Gehring and Palka [54] as a tool for the study of quasiconformal homogeneity. It turns out that the quasihyperbolic metric is invariant under Euclidean similarities, but it is not invariant under conformal mappings, not even under Möbius transformations. From Gehring and Osgood's result [53] it follows that for each domain $\Omega \subseteq \mathbb{R}^n$ and points $x, y \in \Omega$, there exists a quasihyperbolic geodesic and that the quasihyperbolic metric is quasiinvariant under quasiconformal mappings. Another hyperbolic type metric of interest is the distance ratio metric j_G. The hyperbolic metric in the unit ball or half space is Möbius invariant. However, the distance ratio metric is not invariant under Möbius transformations. Therefore it is natural to ask what is its Lipschitz constant under conformal mappings or Möbius transformations in higher dimensions. Gehring and Osgood proved that this metric is not changed by more than a factor 2 under Möbius transformations. In Chap. 4 we present a refinement of this result given by Simić and Vuorinen [147]. It turns out that the factor 2 can be improved in the cases of the unit ball and the punctured unit ball in $\mathbb{R}^n$ and that, in fact, the best possible factors can be identified. The Gehring–Osgood theorem provides a Hölder-type estimate for the modulus of continuity of quasiconformal mappings with respect to the quasihyperbolic metrics of the domain and the target domain of the mapping. As shown by Vuorinen in [157, Example 3.10], there is no counterpart of this result for analytic functions in the plane. Recall that the famous Schwarz–Pick lemma provides a Lipschitz-type modulus of continuity estimate for analytic functions of the unit disk into itself with respect to the hyperbolic metrics of the domain and target disks. In recent years there has been a large amount of activity in the study of hyperbolic type metrics [39, 58, 60, 81, 82, 97].

Chapter 5 includes bi-Lipschitz properties of harmonic quasiconformal mappings in the planar and the higher dimensional case. The author has shown in [99] that HQC mappings between any two proper domains in the plane are bi-Lipschitz with respect to the corresponding quasihyperbolic metrics. In the course of proving this, the author also showed that a sense preserving harmonic mapping between two planar domains has a superharmonic logarithm of the Jacobian. This theorem has found some applications on its own. For instance, Tadeusz Iwaniec [70] used this result in establishing the minimum principle for the Jacobian determinant, a remarkable novelty which leads us to the new analytic proof of the celebrated Radó–Kneser–Choquet theorem. The result from author's paper [99] was also used for higher dimensional generalizations of the Pavlović's bi-Lipschitz condition. In the joint paper [21] of K. Astala and the author the bi-Lipschitz property is proved for gradient harmonic quasiconformal mappings in the unit ball $\mathbb{B}^3$. However, in higher dimensions, Pavlović's approach seems difficult to apply. The Lipschitz property follows from the regularity theory of elliptic PDEs established by Kalaj [76] and by a simple and self-contained argument that works for all dimensions given in [21]. The co-Lipschitz condition is much more difficult to tackle, as it is not even known to hold in higher dimensions when the HQC mappings have nonvanishing Jacobian. Indeed, a famous example by J. C. Wood [162] shows that Jacobian can vanish for harmonic injective mappings in dimensions higher than two.

In Chap. 6 we present a result of the author from [87] which solves a problem posed by Pavlović about the functions that are quasi-nearly subharmonic (QNS) in the plane by showing that this class is conformally invariant. An analogous result for regularly oscillating (RO) functions is also proved in the same paper. These results motivated Riihentaus (who introduced the QNS class) and Dovgoshey to partially extend these results to the class of bi-Lipschitz mappings [37]. Since bi-Lipschitz mappings are quasiconformal, the general problem of invariance of QNS and RO classes remained open. In cooperation with P. Koskela, the author solved this problem in [90] by showing that both classes remain invariant under quasiregular mappings with bounded multiplicity, which includes the quasiconformal case. In the paper [90] the problem of composition $u \circ \phi$ is solved, where u is QNS and ϕ is QC, not only in the plane but also in the space. This generalizes results from [87], as well as results of Riihentaus and Dovgoshey which were based on [87].

In Chap. 7 we introduce problems related to characterization of boundary values of harmonic quasiconformal mappings. From many aspects for harmonic quasiconformal mappings, the quantity $\log J(z, f)$ seems the natural counterpart of $\log f'(z)$. In particular, the question arises if the counterparts of Pommerenke's and Kellogs' theorems hold for HQC mappings and $\log J(z, f)$ instead of conformal mappings and $\log f'(z)$. In the last chapter we also pose some problems in this direction.

Chapter 2
Quasiconformal and Quasiregular Harmonic Mappings

In this chapter we build the foundation for the work that comes in the rest of the book. We begin with the definition of two conformal invariants, the modulus of a curve family and the capacity of a condenser, which are two closely related notions. These tools enable us to define quasiconformal and quasiregular mappings which are the basic classes of mappings to be studied. Several examples of quasiconformal mappings are given illustrating the importance of this class of functions and their role in Geometric Function Theory. Moduli of continuity of harmonic mappings, which are either quasiconformal or quasiregular at the same time, are considered and some sharp estimates are given for all dimensions $n \geq 2$. In particular, we study the case of Lipschitz continuity of mappings defined in the unit ball.

2.1 Moduli of Curve Families

Conformal invariance has played a predominant role in the Geometric Function Theory during the past century. The set of landmark works include the pioneering contributions of Grötzsch and Teichmüller prior to the Second World War and the paper of Ahlfors and Beurling [6] in 1950. These results led to far-reaching applications and have stimulated many later studies [92]. Gehring and Väisälä [51, 155] built the theory of quasiconformal mappings in $\mathbb{R}^n$ based on the notion of the modulus of a curve family introduced by Ahlfors and Beurling [6] in the plane and extended to $\mathbb{R}^n$ by Fuglede [45], which is an essential tool in the investigation of quasiconformal mappings and the theory of quasiregular mappings (see [133]).

For the notions of path integral and modulus $M(\Gamma)$ of a family Γ of curves in $\mathbb{R}^n$, we refer the reader to [155] and [158] where the following result is found.

Theorem 2.1 ([155, Theorem 1.3, p. 2]) *The length function* $s : [a, b] \longrightarrow \mathbb{R}$ *of rectifiable path* $\alpha : [a, b] \longrightarrow \mathbb{R}^n$ *has the following properties:*

© Springer Nature Switzerland AG 2019
V. Todorčević, *Harmonic Quasiconformal Mappings and Hyperbolic Type Metrics*,
https://doi.org/10.1007/978-3-030-22591-9_2

1. $a \le t_1 \le t_1 \le b$ implies $l(\alpha|_{[t_1,t_2]}) = s(t_2) - s(t_1) \ge |\alpha(t_2) - \alpha(t_1)|$,
2. s is increasing,
3. s is continuous,
4. s is absolutely continuous iff α is absolutely continuous,
5. $s'(t)$ and $\alpha'(t)$ exist a.e. and $s'(t) = |\alpha'(t)|$ a.e.,
6. $l(\alpha) \ge \int_a^b s'(t)\,dt = \int_a^b |\alpha'(t)|\,dt$,

where the equality holds iff s (or α) is absolutely continuous.

For locally rectifiable path $\gamma : \Delta \longrightarrow X$ and a continuous function $f : \gamma\Delta \longrightarrow$ $[0, \infty]$, the path integral is defined in two steps. Recall that γ^o is the normal representation of a rectifiable path γ.

1. If γ is rectifiable, we set

$$\int_\gamma f\,ds = \int_0^{l(\gamma)} f(\gamma^o(t))|(\gamma^o)'(t)|\,dt.$$

2. If γ is locally rectifiable, we set

$$\int_\gamma f\,ds = \sup\left\{ \int_\beta f\,ds \ : \ l(\beta) < \infty,\ \beta \text{ is subpath of } \gamma \right\}.$$

Let us consider a family Γ of curves in $\overline{\mathbb{R}^n}$. We say that a nonnegative Borel measurable $\rho : \mathbb{R}^n \to \overline{\mathbb{R}}$ is an *admissible metric* for Γ if

$$l_\rho(\gamma) = \int_\gamma \rho\,ds \geqslant 1 \text{ for each locally rectifiable } \gamma \in \Gamma.$$

Let $F(\Gamma)$ be the set of all admissible metrics for Γ. Finally, for each $p \geqslant 1$ we define the *p-modulus* of Γ by

$$M_p(\Gamma) = \inf_{\rho \in F(\Gamma)} \int_{\mathbb{R}^n} \rho^p dm.$$

If $F(\Gamma) = \emptyset$, we define $M_p(\Gamma) = \infty$. This happens only if Γ contains a constant path[1] because otherwise the constant function $\rho(x) = \infty$ belongs to $F(\Gamma)$. Clearly, $0 \le M_p(\Gamma) \le \infty$.

The case $p = n$ is the most important. In that case we simply write $M(\Gamma)$ for the modulus.

Note: Let $\Gamma_{lr} = \{\gamma \in \Gamma \mid \gamma \text{ is locally rectifiable}\}$ and $\Gamma_r = \{\gamma \in \Gamma \mid \gamma \text{ is rectifiable}\}$. By definition, we see that $M_p(\Gamma) = M_p(\Gamma_{lr})$. It can be shown that $M(\Gamma) = M(\Gamma_r)$.

---

[1] A case which will never be considered in this book.

Theorem 2.2 ([155, Theorem 6.2, p. 16]) *M_p is an outer measure on the set of all curves in $\overline{\mathbb{R}}^n$:*

1. $M_p(\emptyset) = 0$,
2. $\Gamma_1 \subset \Gamma_2 \Rightarrow M_p(\Gamma_1) \leqslant M_p(\Gamma_2)$,
3. $M_p\left(\bigcup_{i=1}^\infty \Gamma_i\right) \leq \sum_{i=1}^\infty M_p(\Gamma_i)$.

Proof

1. Since the zero function belongs to $F(\emptyset)$, $M_p(\emptyset) = 0$.
2. If $\Gamma_1 \subset \Gamma_2$ then $F(\Gamma_1) \supset F(\Gamma_2)$, whence $M_p(\Gamma_1) \leqslant M_p(\Gamma_2)$.
3. We may assume that $M_p(\Gamma_i) < \infty$ for each i. For $\varepsilon > 0$ pick $\rho_i \in F(\Gamma_i)$ such that

$$\int \rho_i^p \, dm < M_p(\Gamma_i) + \frac{\varepsilon}{2^i}.$$

Then the function $\rho = \left(\sum \rho_i^p\right)^{1/p} \geqslant \rho_i$ is admissible for each family Γ_i, and therefore for $\bigcup_i \Gamma_i$. Thus

$$M_p(\Gamma) \leqslant \int_{\mathbb{R}^n} \rho^p \, dm = \int_{\mathbb{R}^n} \sum_i \rho_i^p \, dm = \sum_i \int_{\mathbb{R}^n} \rho_i^p \, dm \leqslant \sum_i M_p(\Gamma_i) + \varepsilon.$$

Letting $\varepsilon \to 0$ we obtain

$$M_p\left(\bigcup_i \Gamma_i\right) \leqslant \sum_i M_p(\Gamma_i).$$

$\square$

Definition 2.3 ([155, Definition 6.3, p. 17]) Let Γ_1 and Γ_2 be curve families in $\mathbb{R}^n$. We say that Γ_2 is *minorized* by Γ_1 and denote $\Gamma_2 > \Gamma_1$ if every $\gamma \in \Gamma_2$ has a subcurve which belongs to Γ_1.

Theorem 2.4 ([155, Theorem 6.4, p. 17]) *If $\Gamma_1 < \Gamma_2$, then $M_p(\Gamma_1) \geqslant M_p(\Gamma_2)$.*

Proof If ρ is admissible for Γ_1, then it is also admissible for Γ_2, and hence $F(\Gamma_1) \subset F(\Gamma_2)$. Therefore $M_p(\Gamma_1) \geqslant M_p(\Gamma_2)$. $\square$

Note that if $\Gamma_1 \supset \Gamma_2$, then $\Gamma_1 < \Gamma_2$. So, Theorem 2.2, part 2, is a special case of Theorem 2.4.

Roughly speaking, $M_p(\Gamma)$ is large if the family Γ is large or if curves in Γ are short.

Definition 2.5 ([155, Definition 6.6, p. 17]) The curve families $\Gamma_1, \Gamma_2, \ldots$ are called *separate* if there exist disjoint Borel sets E_i in $\mathbb{R}^n$ such that if $\gamma \in \Gamma_i$ is locally rectifiable, then $\int_\gamma g_i \, ds = 0$, where g_i is the characteristic function of E_i^c.

This condition says that each locally rectifiable $\gamma \in \Gamma_i$ lies almost entirely (in terms of arc length) in E_i.

Theorem 2.6 ([155, Theorem 6.7, p. 18]) *If $\Gamma_1, \Gamma_2, \ldots$ are separate and if $\Gamma < \Gamma_i$ for all i, then*

$$M_p(\Gamma) \geqslant \sum_i M_p(\Gamma_i).$$

Proof For $\rho \in F(\Gamma)$ set $\rho_i(x) = (1 - g_i(x)) \rho(x)$. Then $\rho_i \in F(\Gamma_i)$, whence

$$\sum_i M_p(\Gamma_i) \leqslant \sum_i \int \rho_i^p dm = \sum_i \int_{E_i} \rho^p dm \leqslant \int \rho^p dm.$$

Thus $\sum_i M_p(\Gamma_i) \leqslant M_p(\Gamma)$.   □

Note that if $\Gamma = \bigcup \Gamma_i$, where Γ_i are separated, then by combining this theorem with subadditivity (Theorem 2.2, part 3), one gets $\sum_i M_p(\Gamma_i) = M_p(\Gamma)$.

If we apply this to a single Γ_1, we obtain the previous theorem.

It is in general very difficult to compute $M_p(\Gamma)$ for a given family Γ. However, choosing a suitable $\rho \in F(\Gamma)$ one can often get a good upper bound for $M_p(\Gamma)$, for if we take any $\rho \in F(\Gamma)$, then $M_p(\Gamma) \leqslant \int \rho^p dm$.

Theorem 2.7 ([155, Theorem 7.1, p. 20]) *Suppose that the curves of a family Γ lie in a Borel set $G \subset \mathbb{R}^n$, and that $l(\gamma) \geqslant r > 0$ for every locally rectifiable $\gamma \in \Gamma$. Then*

$$M_p(\Gamma) \leqslant \frac{m(G)}{r^p}.$$

Proof Define $\rho : \mathbb{R}^n \to \mathbb{R}$ by

$$\rho(x) = \begin{cases} \dfrac{1}{r}, & x \in G, \\[2mm] 0, & x \notin G. \end{cases}$$

Then $\int_\gamma \frac{1}{r} ds \geqslant r \cdot \frac{1}{r} = 1$, so $\rho \in F(\gamma)$ and

$$\int_{\mathbb{R}^n} \rho^p dm = \frac{1}{r^p} \int_G dm = \frac{m(G)}{r^p}.$$

                                                                                        □

To get a good lower bound for $M_p(\Gamma)$ is much more difficult. Now one has to prove an estimate for each $\rho \in F(\Gamma)$, in contrast to the opposite inequality where a choice of a single ρ is sufficient. Usually that estimate is done by combining the Hölder's inequality and the Fubini's theorem.

Fig. 2.1 Cylinder

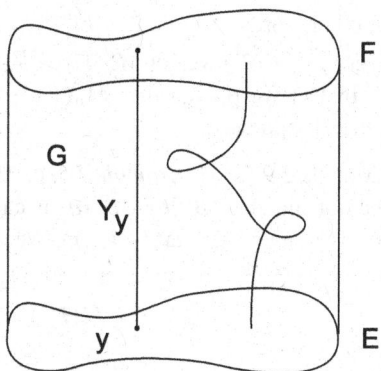

We introduce notation:

For $E, F, G \subset \overline{\mathbb{R}}^n$ let $\Delta(E, F, G)$ be the family of all closed curves joining E to F within G. More precisely, a path $\gamma : [a, b] \to \overline{\mathbb{R}}^n$ belongs $\Delta(E, F, G)$ iff $\gamma(a) \in E$, $\gamma(b) \in F$ and $\gamma(t) \in G$ for $a < t < b$.

Example 2.8 ([155, Example 7.2, p. 21] (Fig. 2.1) (The Cylinder Example)) Let E be a Borel set in $\mathbb{R}^{n-1}$ and let $h > 0$. Set

$$G = \{x \in \mathbb{R}^n \mid (x_1, \ldots, x_{n-1}) \in E \text{ and } 0 < x_n < h\}.$$

Then G is a cylinder with bases E and $F = E + h\,e_n$ and with height h. Set $\Gamma = \Delta(E, F, G)$. We show that

$$M_p(\Gamma) = \frac{m_{n-1}(E)}{h^{p-1}} = \frac{m(G)}{h^p}.$$

Since $l(\gamma) \geqslant h$ for every $\gamma \in \Gamma$, Theorem 2.7 implies $M_p(\Gamma) \geqslant \frac{m(G)}{h^p}$.

Let $\rho \in F(\Gamma)$. For each $y \in E$ let $Y_y : [0, h] \to \mathbb{R}^n$ be the vertical segment $Y_y(t) = y + t\,e_n$. Then $Y_y \in \Gamma$. Assume that $p > 1$. Then we have

$$1 \leqslant \left(\int_{Y_y} 1 \cdot \rho \, ds \right)^p \leqslant \left[\left(\int_0^h 1^q \right)^{1/q} \left(\int_0^h \rho^p \, ds \right)^{1/p} \right]^p$$

$$= h^{p/q} \int_0^h \rho^p \, ds = h^{p-1} \int_0^h \rho^p(y + t\,e_n) \, dt.$$

Here we used admissibility of ρ in the first and Hölder inequality in the second inequality. Now we integrate over E and use Fubini's theorem:

$$\int_E 1 \, dm \leqslant \int_E h^{p-1} dm_{n-1} \int_0^h \rho^p(y + t\,e_n) \, dt,$$

in other words $m_{n-1}(E) \leqslant h^{p-1} \int_G \rho^p dm \leqslant h^{p-1} \int \rho^p dm$. Since this holds for every $\rho \in F(\Gamma)$ we get $M_p(\Gamma) \geqslant \frac{m_{n-1}(E)}{h^{p-1}}$.

In fact we proved that $M_p(\Gamma) = M_p(\Gamma_0)$, where Γ_0 is the subfamily of all vertical segments.

Example 2.9 ([155, Example 7.5, p. 22] (The Spherical Ring))   If $0 < a < b < \infty$, the domain $A = B^n(b) \setminus \overline{B}^n(a)$ is called a spherical ring. Let $E = S^{n-1}(a)$, $F = S^{n-1}(b)$, and $\Gamma_A = \Delta(E, F, A)$. We are going to prove that

$$M(\Gamma_A) = \omega_{n-1}(\log \frac{b}{a})^{1-n},$$

where ω_{n-1} is the surface measure of the unit sphere S^{n-1}. Pick $\rho \in F(\Gamma_A)$. For each unit vector $y \in S^{n-1}$ we let $Y_y : [a, b] \to \mathbb{R}^n$ be the radial segment, defined by $Y_y(t) = t \cdot y$. By Hölder's inequality (for $p = \frac{n}{n-1}$ and $q = n$) we obtain

$$1 \leqslant \left(\int_{Y_y} \rho \, ds \right)^n = \left(\int_a^b (\rho(ty) \cdot t^{1-\frac{1}{n}}) t^{\frac{1}{n}-1} dt \right)^n$$

$$\leqslant \left[\left(\int_a^b (\rho(ty) \cdot t^{1-\frac{1}{n}})^n dt \right)^{1/n} \left(\int_a^b (t^{\frac{1}{n}-1})^{\frac{n}{n-1}} dt \right)^{(n-1)/n} \right]^n$$

$$= \int_a^b \rho^n(ty) t^{n-1} dt \cdot \left(\int_a^b t^{-1} dt \right)^{n-1} = (\log \frac{b}{a})^{n-1} \int_a^b \rho^n(ty) \, t^{n-1} dt.$$

Integrating over $y \in S^{n-1}$ yields

$$\omega_{n-1} \leqslant (\log \frac{b}{a})^{n-1} \int \rho^n dm.$$

Since it is true for each $\rho \in F(\Gamma)$, we get $M(\Gamma_A) \geqslant \omega_{n-1}(\log \frac{b}{a})^{1-n}$. Now we choose a concrete $\rho \in F(\Gamma)$:

$$\rho(x) = \begin{cases} \dfrac{1}{|x| \log \frac{b}{a}}, & x \in A, \\[2ex] 0, & x \notin A. \end{cases}$$

It is geometrically obvious that $l_\rho(\gamma) \geqslant \frac{1}{\log \frac{b}{a}} \int_a^b \frac{1}{r} dr = 1$, so $\rho \in F(\Gamma)$.

Note that this can be generalized to cone domains. Let Y be a Borel set in S^{n-1} and let C be the cone $\{x \in \mathbb{R}^n \setminus \{0\} \, | \, x/|x| \in Y\}$. Set $\Gamma = \{\gamma \in \Gamma_A \, | \, |\gamma| \subset C\}$ where A is as above. Then we have

$$M(\Gamma) = m_{n-1}(Y)(\log \frac{b}{a})^{1-n}.$$

Example 2.10 ([155, Example 7.8, p. 23] (The degenerated ring)) Let $\Gamma = \Delta(E, F, G)$ where $E = \{0\}$, $F = S^{n-1}(b)$, and $G = B^n(b) \setminus \{0\}$. Since $\Gamma > \Gamma_A$ for every spherical ring $A = B^n(b) \setminus \overline{B}^n(a)$, we obtain (from Theorem 2.4 and Example 2.9) $M(\Gamma) \leqslant M(\Gamma_1) = \omega_{n-1}(\log \frac{b}{a})^{1-n}$. Since this is true for every $a > 0$ when $a \rightarrow 0$, we get $M(\Gamma) = 0$.

Example 2.11 ([155, Example 7.9, p. 23] (Paths Through a Point)) Let Γ be the family of all nonconstant paths γ passing through a fixed point $a \in \mathbb{R}^n$. We prove that $M(\Gamma) = 0$. Without loss of generality assume $a = 0$ and set $\Gamma_k = \Gamma_A$, where $A = \{x \in \mathbb{R}^n \mid 0 < |x| < \frac{1}{k}\}$. Since $\Gamma > \bigcup_k \Gamma_k$, then $M(\Gamma) \leqslant \sum_k M(\Gamma_k) = 0$.

Suppose that A is a subset of $\overline{\mathbb{R}^n}$ and that $f : A \rightarrow \overline{\mathbb{R}^n}$ is continuous. If Γ is a family of curves in A, then the family $\Gamma' = \{f \circ \Gamma \mid \gamma \in \Gamma\}$ is called the *image* of Γ under f.

Theorem 2.12 ([155, Theorem 8.1, p. 25]) *If $f : D \rightarrow D'$ is conformal, then $M(\Gamma') = M(\Gamma)$ for every family of curves Γ in D.*

Proof We can assume that Γ and Γ' do not pass through ∞. Let $\rho \in F(\Gamma)$, and set

$$\rho_1(f(x)) = \rho(x) \cdot \frac{1}{|f'(x)|}, \qquad \text{for } x \in D \text{ and } 0 \text{ otherwise.}$$

Since, $\rho \in F(\Gamma)$, we have

$$\int_\gamma \rho \, |dx| \geqslant 1, \qquad \text{for } \gamma \in \Gamma.$$

A change of variables gives (see [155, Theorem 5.6, p. 14])

$$y = f(x), \qquad |dy| = |f'(x)| \cdot |dx|$$

and hence

$$\int_{f(\gamma)} \rho_1(y) \, |dy| = \int_\gamma \rho_1(f(x)) \, |f'(x)| \, |dx| = \int_\gamma \rho(x) \, |dx| \geqslant 1,$$

so $\rho_1 \in F(\Gamma')$. We have

$$M(\Gamma') = \inf_{\rho_1} \int \rho_1^n(y) \, dy = \inf_\rho \int \rho^n(x) \frac{1}{|f'(x)|^n} J_f(x) \, dx = M(\Gamma).$$

$\square$

Note that this works only for $p = n$ (cancellation of $\frac{1}{|f'(x)|^p}$ with $J_f(x)$ occurs only for $p = n$). However, if $f(x) = k\,x$, $(k > 0)$, then by a similar calculation one gets the following result.

Theorem 2.13 ([155, Theorem 8.2, p. 25]) $M_p(k\,\Gamma) = k^{n-p}M_p(\Gamma)$.

Theorem 2.14 ([155, Theorem 10.12, p. 31]) *Suppose that* $0 < a < b$ *and that E and F are disjoint sets such that every sphere* $S^{n-1}(t)$, $a < t < b$, *meets both E and F. If G contains the spherical ring* $A = B^n(b) \setminus B^n(a)$ *and if* $\Gamma = \Delta(E, F, G)$, *then*

$$M(\Gamma) \geqslant c_n \log \frac{b}{a}, \tag{2.1}$$

where

$$c_n = \frac{1}{2}\omega_{n-2}\left(\int_0^\infty t^{-\frac{n-2}{n-1}}(1+t^2)^{-\frac{1}{n-1}}dt\right)^{1-n}, \quad c_2 = \frac{1}{\pi}.$$

There is equality in (2.1) if $G = A$ and if E and F are the components of $L \cap A$, where L is a line through the origin.

It is important to observe that the constants Ω_n and ω_{n-1} depend strongly on the dimension, indeed $\Omega_n \to 0$ and $\omega_{n-1} \to 0$ when $n \to \infty$ by [10, 2.28]. Various estimates for the constant c_n in 2.1 are given [10, pp. 41–44, 458].

Definition 2.15 ([155, p. 32]) A domain $A \subset \overline{\mathbb{R}^n}$ is a *ring* if $C(A)$ has exactly two components, where $C(A)$ denotes the complement of $A \subset \mathbb{R}^n$.

If the components of $C(A)$ are C_0 and C_1, we denote $A = R(C_0, C_1)$, $B_0 = C_0 \cap \overline{A}$ and $B_1 = C_1 \cap \overline{A}$. To each ring $A = R(C_0, C_1)$, we associate the curve family $\Gamma_A = \Delta(B_0, B_1, A)$.

Definition 2.16 ([158]) The modulus $mod(R)$ of the ring $R(C_0, C_1)$ is defined by

$$mod(R) = mod(R(C_0, C_1)) = \left(\frac{\omega_{n-1}}{M(\Delta(C_0, C_1))}\right)^{1/(n-1)}.$$

The capacity $cap(R)$ of $R(C_0, C_1)$ is $M(\Delta(C_0, C_1))$.

The complementary components of the Grötzsch ring $R_{G,n}(s)$ in $\mathbb{R}^n$ are $\overline{\mathbb{B}}^n$ and $[s \cdot e_1, \infty]$, $s > 1$, while those of the Teichmüller ring $R_{T,n}(t)$ are $[-e_1, 0]$ and $[t\,e_1, \infty]$, $t > 0$. We shall need two special functions $\gamma_n(s)$, $s > 1$ and $\tau_n(t)$, $t > 0$ to designate the moduli of the families of all those curves which connect the complementary components of the Grötzsch and Teichmüller rings in $\mathbb{R}^n$ respectively (Fig. 2.2).

$$\gamma_n(s) = M(\Gamma_s) = \gamma(s), \quad \Gamma_s = \Gamma_{R_{G,n}}(s),$$

$$\tau_n(t) = M(\Delta_t) = \tau(t), \quad \Delta_t = \Gamma_{R_{T,n}}(t).$$

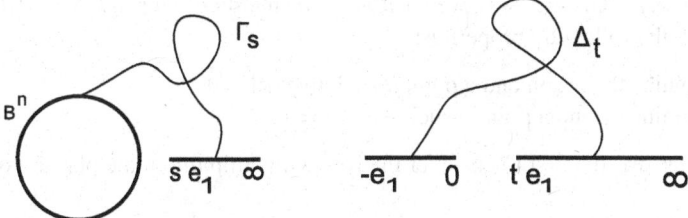

Fig. 2.2 The Grötzsch and Teichmüller rings

These functions are related by a functional identity [48, Lemma 6]

$$\gamma_n(s) = 2^{n-1}\tau_n(s^2 - 1).\tag{2.2}$$

We define the functions $\Phi = \Phi_n$ and $\Psi = \Psi_n$ by

$$\omega_{n-1}(\log(\Phi(s)))^{1-n} = \gamma_n(s), \quad s > 1$$

$$\omega_{n-1}(\log(\Psi(t)))^{1-n} = \tau_n(t), \quad t > 0.$$

Lemma 2.17 ([158, 7.20, p. 88]) *The function $\Phi(t)/t$ is increasing for $t > 1$ and $\Psi(t - 1) = \Phi(\sqrt{t})^2$ for $t > 1$. Moreover, the functions γ_n and τ_n are strictly decreasing.*

By the previous lemma, the function $\log \Phi(t) - \log t$ is increasing and therefore has a limit as $t \to \infty$. We define the number λ_n by

$$\log \lambda_n = \lim_{t \to \infty} (\log \Phi(t) - \log t).$$

This number is usually called the Grötzsch (ring) constant. Only for $n = 2$ is the exact value of the Grötzsch constant known, $\lambda_2 = 4$. Various estimates for λ_n, $n \geq 3$, are given in [48, p. 518], [33, pp. 239–241], [8].

Lemma 2.18 ([158, 7.22, p. 89]) *For each $n \geq 2$ there exists a number $\lambda_n \in [4, 2 \cdot e^{n-1})$, $\lambda_2 = 4$ such that*

1. $t \leq \Phi(t) \leq \lambda_n t, t > 1$,
2. $t + 1 \leq \Psi(t) \leq \lambda_n^2(t + 1), t > 0$.

Furthermore, $\lambda_n^{1/n} \to e$ as $n \to \infty$ and in particular, $\lambda_n \to \infty$ as $n \to \infty$.
The inequality $\Phi(t) \leq \lambda_n t$ is equivalent, by the definition of the function Φ, to

$$\gamma_n(t) \geq \omega_{n-1}(\log \lambda_n t)^{1-n}, \quad t > 1.\tag{2.3}$$

Definition 2.19 Given $r > 0$, we let $R\Psi_n(r)$ be the set of all rings $A = R(C_0, C_1)$ in $\overline{\mathbb{R}^n}$ with the following properties:

1. C_0 contains the origin and a point a such that $|a| = 1$.
2. C_1 contains ∞ and a point b such that $|b| = r$.

Teichmüller was the first to consider the following infimum in the planar case ($n = 2$),

$$\tau_n(r) = \inf M(\Gamma_A),$$

where infimum is taken over all rings $A \in R\Psi_n(r)$. For $n \geqslant 3$ it was studied in [48].

Theorem 2.20 ([155, Theorem 11.7, pp. 34–35], [48]) *The function* $\tau_n :$ $(0, \infty) \to (0, \infty)$ *has the following properties:*

1. τ_n *is decreasing,*
2. $\lim_{r \to \infty} \tau_n(r) = 0$,
3. $\lim_{r \to 0} \tau_n(r) = \infty$,
4. $\tau_n(r) > 0$ *for every* $r > 0$.

Moreover, $\tau_n : (0, \infty) \to (0, \infty)$ *and* $\gamma_n : (1, \infty) \to (0, \infty)$ *are homeomorphisms.*

From the conformal invariance of the modulus and from the definition of τ_n, we obtain the following estimate:

Theorem 2.21 ([155, Theorem 11.9, p. 36], [48]) *Suppose that* $A = R(C_0, C_1)$ *is a ring and that* $a, b \in C_0$ *and* $c, \infty \in C_1$. *Then*

$$M(\Gamma_A) \geqslant \tau_n \left(\frac{|c - a|}{|b - a|} \right).$$

2.2 Definition of Quasiconformal Mappings

Let $f : D \to D'$ be a homeomorphism. If Γ is a family of curves in D, then Γ' denotes the family $\{f \circ \gamma \mid \gamma \in \Gamma\}$ of curves in D'. We set

$$K_I(f) = \sup \frac{M(\Gamma')}{M(\Gamma)}, \quad K_O(f) = \sup \frac{M(\Gamma)}{M(\Gamma')},$$

where the suprema are taken over all families of curves $\Gamma \subset D$ such that $M(\Gamma)$ and $M(\Gamma')$ are not simultaneously 0 or ∞.

Note that the two quantities are equal if f is a conformal mapping (see Theorem 2.12).

Definition 2.22 ([155, Definition 13.1, p. 42]) If $f : D \to D'$ is a homeomorphism, $K_I(f)$ is the inner dilatation and $K_O(f)$ is the outer dilatation of f. The maximal dilatation of f is $K(f) = \max\{K_I(f), K_O(f)\}$. If $K(f) \leqslant K < \infty$, f is K-quasiconformal. f is quasiconformal (qc) if $K(f) < \infty$.

Equivalently, f is K-quasiconformal iff

$$\frac{M(\Gamma)}{K} \leqslant M(\Gamma') \leqslant K\, M(\Gamma),$$

for every family of curves Γ in D. This is the geometric definition of quasiconformal mappings.

Definition 2.23 ([155, Definition 26.2, p. 88]) Let $Q = \{x \in \mathbb{R}^n : a_i \leq x_i \leq b_i\}$ be a closed n-interval. A mapping $f : Q \longrightarrow \mathbb{R}^m$ is said to be *ACL* (absolutely continuous on lines) if f is continuous and if f is absolutely continuous on almost every line segment in Q, parallel to the coordinate axes.

Definition 2.24 [155, Definition 26.5, p. 89] An *ACL*-mapping $f : U \longrightarrow \mathbb{R}^m$ is said to be ACL^p, $p \geq 1$, if the partial derivatives of f are locally L^p-integrable. A homeomorphism $f : D \longrightarrow D'$ is ACL^p if the restriction of f to $D\setminus\{\infty, f^{-1}(\infty)\}$ is ACL^p.

Definition 2.25 ([158, Definition 10.1, pp. 127–128], [113]) Let $G \subseteq \mathbb{R}^n$ be a domain. A mapping $f : G \longrightarrow \mathbb{R}^n$ is said to be *quasiregular* (QR) if f is ACL^n and if there exists a constant $K \geq 1$ such that

$$|f'(x)|^n \leq K\, J_f(x), \quad |f'(x)| = \max_{|h|=1} |f'(x)h|, \quad \text{a.e. in } G.$$

Here $f'(x)$ denotes the formal derivative of f at x. The smallest $K \geq 1$ for which this inequality is true is called the *outer dilatation* of f and denoted by $K_O(f)$. If f is quasiregular, then the smallest $K \geq 1$ for which the inequality

$$J_f(x) \leq K\, l(f'(x))^n, \quad l(f'(x)) = \min_{|h|=1} |f'(x)h|,$$

holds a.e. in G is called the *inner dilatation* of f and denoted by $K_I(f)$. The *maximal dilatation* of f is the number $K(f) = \max\{K_I(f), K_O(f)\}$. If $K(f) \leq K$, f is said to be K-quasiregular (K-QR). If f is not quasiregular, we set $K_O(f) = K_I(f) = K(f) = \infty$.

It turns out that injective quasiregular mappings are quasiconformal and that all quasiregular mappings are orientation-preserving ($J_f \geq 0$ a.e). Definition 2.22 gives that a quasiconformal mapping is orientation reversing. Therefore, in particular, all Möbius transformations and all sense-reversing transformations are 1-quasiconformal, but they are not 1-quasiregular because quasiregular mappings are required to have Jacobian greater than zero, hence they are sense-preserving.

Theorem 2.26 ([130, Theorem 1.1]) *For a given $n \geq 3$ there exists a constant $e(n)$ depending on n such that if $f : \Omega \to \mathbb{R}^n$ is a quasiregular mapping of a domain $\Omega \subset \mathbb{R}^n$ and if $K(f) \leq 1 + e(n)$, then f is a local homeomorphism.*

Note that from this theorem we have that for small K, harmonic K-quasiregular mappings are diffeomorphism and, therefore, locally quasiconformal (see [155, p. 46]).

The following theorem states simple properties of the dilatations just introduced above relative the compositions and inverses and its proof is left to the reader.

Theorem 2.27 ([155, Theorem 13.2, p. 42])

1. $K_I(f^{-1}) = K_O(f)$,
2. $K_O(f^{-1}) = K_I(f)$,
3. $K(f^{-1}) = K(f)$,
4. $K_I(f \circ g) \leqslant K_I(f)K_I(g)$,
5. $K_O(f \circ g) \leqslant K_O(f)K_O(g)$,
6. $K(f \circ g) \leqslant K(f)K(g)$.

Corollary 2.28 ([155, Corollary 13.3, p. 42]) *If f is K-quasiconformal, then f^{-1} is K-quasiconformal.*

Proof This follows from Theorem 2.27, part 3.   □

Corollary 2.29 *If $h = f \circ g$, where f is K_1-quasiconformal and g is K_2-quasiconformal, then h is K_1K_2-quasiconformal.*

Proof This is a consequence of Theorem 2.27, part 6.   □

Definition 2.30 ([155, Definition 14.1, p. 43]) Let $A : \mathbb{R}^n \to \mathbb{R}^n$ be a linear bijection. The numbers

$$H_I(A) = \frac{|\det(A)|}{l^n(A)}, \quad H_O(A) = \frac{|A|^n}{|\det(A)|}, \quad H(A) = \frac{|A|}{l(A)},$$

where $l(A) = \inf_{\|x\|=1} \|Ax\|$, are called the *inner, outer*, and *linear dilatation* of A, respectively.

They have the following geometric interpretation: The image of the unit ball $\mathbb{B}^n$ under A is an ellipsoid $E(A)$. Let $B_I(A)$ and $B_O(A)$ be the inscribed and circumscribed balls of $E(A)$, respectively. Then

$$H_I(A) = \frac{m(E(A))}{m(B_I(A))} = \frac{a_1 \cdots a_{n-1}}{a_n^{n-1}},$$

$$H_O(A) = \frac{m(B_O(A))}{m(E(A))} = \frac{a_1^{n-1}}{a_2 \cdots a_n}, \quad H(A) = \frac{a_1}{a_n},$$

where $a_1 \geqslant a_2 \geqslant \cdots \geqslant a_n$ are the semi-axes of $E(A)$.

Next we turn to a more interesting task: finding conditions on quasiconformality in the case of a C^1 mapping in terms of its derivative. This is an analytic approach to quasiconformal mappings.

Module estimates will play crucial role here. In order to do this, we define

$$H_O(f'(x)) = \frac{|(f'(x))|^n}{|J_f(x)|}, \qquad H_I(f'(x)) = \frac{|J_f(x)|}{l(f'(x))^n},$$

where $J_f(x) \neq 0$ is Jacobian of f.

Theorem 2.31 ([155, Theorem 15.1, p. 46]) *Suppose that* $f : D \to D'$ *is a diffeomorphism. Then*

$$K_I(f) = \sup_{x \in D} H_I(f'(x)), \qquad K_O(f) = \sup_{x \in D} H_O(f'(x)).$$

Theorem 2.32 ([155, Theorem 15.3, p. 47]) *Let* $f : D \to D'$ *be a homeomorphism. If* f *is differentiable at a point* $a \in D$ *and if* $K_O(f) < \infty$, *then*

$$|f'(a)|^n \leqslant K_O(f) \cdot |J_f(a)|.$$

Corollary 2.33 ([155, Corollary 15.5, p. 48]) *A diffeomorphism* $f : D \to D'$ *is* K-*quasiconformal iff the double inequality*

$$\frac{|f'(x)|^n}{K} \leqslant |J_f(x)| \leqslant K \cdot l(f'(x))^n$$

holds for every $x \in D$.

Example 2.34 ([155, Example 16.1, pp. 48–49] (A Linear Mapping)) Let $A : \mathbb{R}^n \to \mathbb{R}^n$ be a linear bijection. Then $A'(x) = A$ for all $x \in \mathbb{R}^n$. From Theorem 2.31 we obtain

$$K_I(A) = H_I(A) \qquad K_O(A) = H_O(A).$$

Thus A is quasiconformal.

Example 2.35 ([155, Example 16.2, p. 49] (A Radial Mapping)) Let $a \neq 0$ be a real number, and set $f(x) = |x|^{a-1}x$. Then f is a diffeomorphism of $\mathbb{R}^n \setminus \{0\}$ onto itself. We can extend f to a homeomorphism $f^* : \overline{\mathbb{R}^n} \longrightarrow \overline{\mathbb{R}^n}$ by defining $f^*(0) = 0$, $f^*(\infty) = \infty$ for $a > 0$ and $f^*(0) = \infty$, $f^*(\infty) = 0$ for $a < 0$. Then f is a quasiconformal with

$$K_I(f) = |a|, \qquad K_O(f) = |a|^{n-1} \qquad \text{if } |a| \geqslant 1,$$
$$K_I(f) = |a|^{1-n}, \qquad K_O(f) = |a|^{-1} \qquad \text{if } |a| \leqslant 1.$$

Example 2.36 ([86], (General Radial Mappings)) Now we consider radial mappings of a more general type:

$$f(x) = \varphi(|x|) \cdot \frac{x}{|x|},$$

Fig. 2.3 Radial mappings

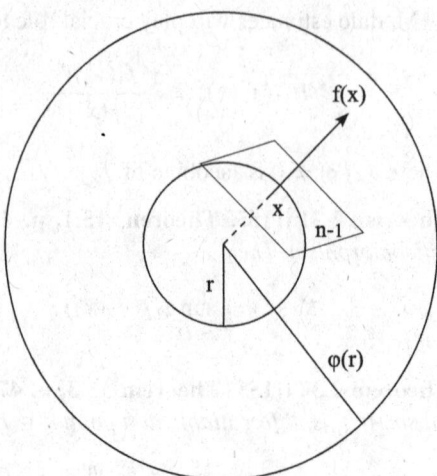

where φ is continuously differentiable on $[0, +\infty)$, $\varphi'(r) > 0$ and $\varphi(0) = 0$ ([86]).
Now we want to calculate the Jacobi matrix of f at a given point x. In fact, we are
interested in $l(f'(x))$ and $|f'(x)|$. It is easier to work in a new rectangular coordinate
system, one coordinate is along the vector x and the other coordinates are in the
tangent plane ($(n-1)$-dimensional hyperplane) to the sphere through x (Fig. 2.3).
Then we have

$$\left[\frac{\partial f}{\partial x}\right]_{i,j=1}^{n} = \begin{bmatrix} \varphi'(r) & & & 0 \\ & \frac{\varphi(r)}{r} & & \\ & & \ddots & \\ 0 & & & \frac{\varphi(r)}{r} \end{bmatrix}.$$

Indeed, the rate of stretching along the first coordinate axis is $\varphi'(r)$ by definition
of f. In the tangent hyperplane we have a similarity transformation by a coefficient
$\frac{\varphi(r)}{r}$, where Δr is "infinitesimal" displacement (Fig. 2.4).
Hence, we have the following equations:

$$J_f(x) = \varphi'(r) \cdot \left(\frac{\varphi(r)}{r}\right)^{n-1} = \det f'(x),$$

$$l(f'(x)) = l\left(\frac{\partial f}{\partial x}\right) = \min\left\{\varphi'(r), \frac{\varphi(r)}{r}\right\},$$

$$|f'(x)| = \max\left\{\varphi'(r), \frac{\varphi(r)}{r}\right\},$$

$$H_O(f'(x)) = \frac{|f'(x)|^n}{J_f(x)} = \frac{\max\left\{\varphi'(r), \frac{\varphi(r)}{r}\right\}^n}{\varphi'(r)\left(\frac{\varphi(r)}{r}\right)^{n-1}}.$$

Fig. 2.4 The stretching rate

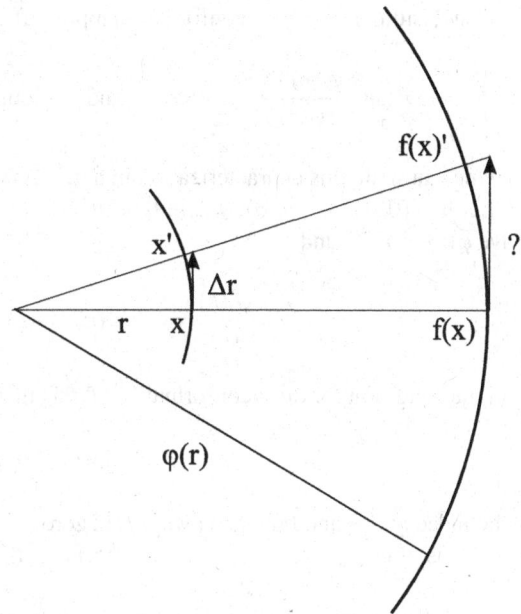

Note that this is bounded iff

$$\sup_{r>0} \frac{\varphi'(r) \cdot r}{\varphi(r)} < +\infty$$

and

$$\sup_{r>0} \frac{\varphi(r)}{r \cdot \varphi'(r)} < +\infty.$$

$$H_I(f'(x)) = \frac{J_f(x)}{l(f'(x)^n} = \frac{\varphi'(r) \left(\frac{\varphi(r)}{r}\right)^{n-1}}{\min\left\{\varphi'(r), \frac{\varphi(r)}{r}\right\}^n}$$

$$= \varphi'(r) \left(\frac{\varphi(r)}{r}\right)^{n-1} \cdot \max\left\{\frac{1}{\varphi'(r)}, \frac{r}{\varphi(r)}\right\}^n.$$

On the other hand, this is bounded iff $\sup_{r>0} \frac{\varphi(r)}{r \cdot \varphi'(r)} < +\infty$ and $\sup_{r>0} \frac{\varphi'(r) \cdot r}{\varphi(r)} < +\infty$, which is the same condition we obtained for H_O.

Conclusion: f is a quasiconformal mapping iff

$$\sup_{r>0} \frac{\varphi'(r) \cdot r}{\varphi(r)} < +\infty \quad \text{and} \quad \sup_{r>0} \frac{\varphi(r)}{r \cdot \varphi'(r)} < +\infty.$$

One can write this characterization in a different form:

Let $\alpha : (0, 1) \to (u, v)$, where $0 < u < v < 1$. For $f(x) = |x|^{\alpha(|x|)-1} x$ we have $\varphi(r) = r^{\alpha(r)}$ and

$$\frac{r\,\varphi'(r)}{\varphi(r)} = r \cdot \alpha'(r) \cdot \ln(r) + \alpha(r)$$

and the condition for quasiconformality of f is that

$$r \cdot \alpha'(r) \cdot \ln(r) + \alpha(r)$$

is bounded above and bounded away from zero.

This is of course true if $\alpha(r) = \alpha \in (0, 1)$. In that case

$$J_f = \begin{bmatrix} \alpha \cdot r^{\alpha-1} & & & 0 \\ & r^{\alpha-1} & & \\ & & \ddots & \\ 0 & & & r^{\alpha-1} \end{bmatrix}.$$

Since $0 < \alpha < 1, l = \alpha \cdot r^{\alpha-1}, |f'(x)| = r^{\alpha-1}, J_f(x) = \alpha \cdot (r^{\alpha-1})^n$, and

$$H_O(f'(x)) = \frac{(r^{\alpha-1})^n}{\alpha \cdot (r^{\alpha-1})^n} = \frac{1}{\alpha}$$

and

$$H_I(f'(x)) = \frac{\alpha \cdot (r^{\alpha-1})^n}{(\alpha \cdot r^{\alpha-1})^n} = \frac{1}{\alpha^{n-1}}.$$

Note that $H_I \geqslant H_O$, and we see that the constant of quasiconformality is $K = \frac{1}{\alpha^{n-1}}$. From here we have $\alpha = K^{1/(1-n)}$. The constant $\alpha = K^{1/(1-n)}$ is the best possible exponent.

Example 2.37 (Hölder Continuity of Radial Mappings [86, Appendix]) In this example we use some results from [31, Lemma 3.3]. Let $f(x) = |x|^{\alpha-1} x, x \in \mathbb{R}^n$, and $\alpha \in (0, 1)$. We prove that f is Hölder continuous with exponent α, i.e.,

$$\forall x, y \in \mathbb{R}^n \qquad |f(x) - f(y)| \leqslant C |x - y|^\alpha$$

with $C = 2^{1-\alpha}$. Note that this Hölder estimate is sharp, since the equality holds for $x = -y$.

We write the function f in the following form:

$$f(x) = |x|^\alpha \frac{x}{|x|}. \tag{2.4}$$

Without loss of generality we can assume that $|x| \geqslant |y| > 0$. Put $k = \frac{|x|}{|y|}$ and $\tilde{x} = \frac{x}{|x|}$, $\tilde{y} = \frac{y}{|y|}$. Then $k \geqslant 1$ and $|\tilde{x}| = |\tilde{y}| = 1$. By the equality (2.4), we need to prove

$$\left| |x|^\alpha \frac{x}{|x|} - |y|^\alpha \frac{y}{|y|} \right| \leqslant C |x - y|^\alpha.$$

By dividing the previous inequality by $|y|^\alpha$, it follows that

$$|k^\alpha \tilde{x} - \tilde{y}| \leqslant C |k \tilde{x} - \tilde{y}|^\alpha.$$

This inequality is equivalent to

$$|k^\alpha \tilde{x} - \tilde{y}|^2 \leqslant C^2 |k \tilde{x} - \tilde{y}|^{2\alpha}.$$

By definition of the inner product, this is equivalent to

$$C^2 \geqslant \frac{k^{2\alpha} - 2k^\alpha \langle \tilde{x}, \tilde{y} \rangle + 1}{(k^2 - 2k \langle \tilde{x}, \tilde{y} \rangle + 1)^\alpha}. \tag{2.5}$$

The inequality

$$C^2 \geqslant \max \frac{k^{2\alpha} - 2k^\alpha \langle \tilde{x}, \tilde{y} \rangle + 1}{(k^2 - 2k \langle \tilde{x}, \tilde{y} \rangle + 1)^\alpha}$$

ensures (2.5). Because $|\langle \tilde{x}, \tilde{y} \rangle| \leqslant 1$, it is sufficient to prove that

$$f(k, t) = \frac{k^{2\alpha} - 2k^\alpha t + 1}{(k^2 - 2kt + 1)^\alpha} \leqslant 2^{2-2\alpha}$$

for each $t \in [-1, 1]$. The function $f(k, t)$ can be written in the form

$$f(k, t) = \frac{(k^\alpha - 1)^2 + 2k^\alpha (1 - t)}{((k - 1)^2 + 2k (1 - t))^\alpha}.$$

By differentiating $f(k, t)$ with respect to t we obtain

$$\frac{\partial f}{\partial t} = \frac{-2k ((k^{\alpha+1} - 1)(1 - k^{\alpha-1}) + (1 - \alpha) ((k^\alpha - 1)^2 + 2k^\alpha(1 - t)))}{(k^2 - 2kt + 1)^{\alpha+1}}$$

$$< 0.$$

Hence, $f(k, t)$ is decreasing as a function of t when $-1 \leqslant t \leqslant 1$. From this we get

$$f(k, t) \leqslant f(k, -1) = \frac{k^{2\alpha} + 2k^{\alpha} + 1}{(k^2 + 2k + 1)^{\alpha}} = \left(\frac{k^{\alpha} + 1}{(k + 1)^{\alpha}} \right)^2.$$

By concavity of the function u^{α},

$$\frac{k^{\alpha} + 1}{(k + 1)^{\alpha}} = \frac{\frac{k^{\alpha} + 1}{2}}{\frac{(k+1)^{\alpha}}{2}} \leqslant \frac{\left(\frac{k+1}{2} \right)^{\alpha}}{\frac{(k+1)^{\alpha}}{2}} = 2^{1-\alpha}.$$

Note that in this example we proved the sharp Hölder estimate for f.

Consider a homeomorphism $f : D \to D'$. Suppose that $x \in D$, $x \neq \infty$ and $f(x) \neq \infty$. For each $r > 0$ such that $S^{n-1}(x, r) \subset D$ we set

$$L(x, f, r) = \max_{|y-x|=r} |f(y) - f(x)|, \qquad l(x, f, r) = \min_{|y-x|=r} |f(y) - f(x)|.$$

$$(2.6)$$

Definition 2.38 ([155, Definition 22.2, p. 78]) The *linear dilatation* of f at x is the number

$$H(x, f) = \limsup_{r \to 0} \frac{L(x, f, r)}{l(x, f, r)}.$$

If $x = \infty$, $f(x) \neq \infty$, we define $H(x, f) = H(0, f \circ u)$ where u is the inversion $u(x) = \frac{x}{|x|^2}$. If $f(x) = \infty$, we define $H(x, f) = H(x, u \circ f)$ (Fig. 2.5).

Example 2.39 The mapping $f : \mathbb{B}^n \to \mathbb{B}^n$, $f(x) = |x|^{\alpha-1}x$ has $H(0, f) = 1$.

This is true for all radial mappings.

Theorem 2.40 ([155, Theorem 22.3, pp. 78–79]) *Suppose that $f : D \to D'$ is a homeomorphism such that one of the following conditions are satisfied for some finite K:*

1. $M(\Gamma_A) \leqslant K \cdot M(\Gamma'_A)$ *for all rings A such that $\overline{A} \subset D$,*
2. $K_O(f) \leqslant K$,
3. $K_I(f) \leqslant K$.

Then $H(x, f)$ is bounded by a constant which depends only on n and K.

Fig. 2.5 A linear dilatation

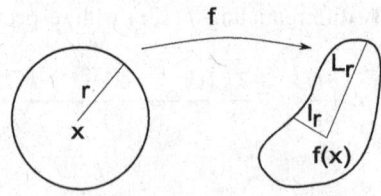

Remark 2.41 Let us explain some history behind the bounds of linear dilatation for $n \geq 2$. First, there was a bound due to Gehring and improved by Vuorinen, see [158, Remark 10.29, p. 136]. A bound which tends to 1 when K tends to 1 was proven by Vuorinen [159] for mappings defined in the whole space in $\mathbb{R}^n$ and this was generalized to the case of mappings defined in subdomains G of $\mathbb{R}^n$ by P. Seittenranta [140], see also [10, 14.37].

2.3 Hölder Continuity of Quasiconformal Mappings

We need more delicate estimates for our further study of Hölder continuity of quasiconformal mappings.

Theorem 2.42 ([158, Theorem 7.47, p. 98]) *For $n \geq 2$, $K > 0$, and $0 \leq r \leq 1$ let*

$$\varphi_K(r) = \varphi_{K,n}(r) = \frac{1}{\gamma_n^{-1}(K\gamma_n(1/r))}. \tag{2.7}$$

We set $\varphi_K(0) = 0$ and $\varphi_K(1) = 1$. Then $\varphi_{K,n} : [0,1] \to [0,1]$ is a homeomorphism and for $K \geq 1$

$$\varphi_K(r) \leq \lambda_n^{1-\alpha} r^\alpha, \quad \alpha = K^{1/(1-n)}, \tag{2.8}$$

$$\varphi_{1/K}(r) \geq \lambda_n^{1-\beta} r^\beta, \quad \beta = K^{1/(n-1)}. \tag{2.9}$$

It should be noticed that in the case $n = 2$, all the functions $\tau_n(t)$, $\gamma_n(t)$, $\varphi_{K,n}(r)$ can be expressed in terms of classical special functions such as complete elliptic integrals while this is not the case for $n \geq 3$ [10]. Therefore, for $n = 2$ one can expect sharper results than for the case of a general dimension $n \geq 3$.

We give here some well-known identities between them that can be found in [10]. First, the function

$$\eta_{K,n}(t) = \tau_n^{-1}(\tau_n(t)/K) = \frac{1 - \varphi_{1/K,n}(1/\sqrt{1+t})^2}{\varphi_{1/K,n}(1/\sqrt{1+t})^2}, \quad K > 0 \tag{2.10}$$

defines an increasing homeomorphism $\eta_{K,n} : (0,\infty) \to (0,\infty)$ (see [10, p.193]). Later, in Theorem 3.24 we will need the constant $(1-a)/a$, where $a = \varphi_{1/K,n}(1/\sqrt{2})^2$, in (3.31) which can be expressed as follows for $K > 1$

$$(1-a)/a = \eta_{K,n}(1) = \tau_n^{-1}(\tau_n(1)/K). \tag{2.11}$$

The following lemma establishes the extremality property of the Grötszch ring and is based on a symmetrization theorem from [48, Theorem 1].

Lemma 2.43 *Let C be a connected compact set contained in the unit disk that contains points 0 and x. Then the capacity of a ring domain with complementary components $C_0 = C$ and $C_1 = \{y : |y| \geqslant 1\}$ is at least $\gamma_n(\frac{1}{|x|})$.*

The next theorem is a counterpart of the Schwarz lemma for quasiconformal mappings.

Theorem 2.44 ([115, Theorem 3.1]) *If $f : \mathbb{B}^n \to \mathbb{B}^n$ is a K-quasiconformal mapping such that $f(0) = 0$, then $|f(x)| \leqslant \varphi_{K,n}(|x|)$ for all $x \in \mathbb{B}^n$, where $\varphi_{K,n}$ is as in Theorem 2.42.*

Proof We fix $x \in \mathbb{B}^n$ (i.e., $|x| < 1$) and consider the ring $R(C_0, C_1)$ where $C_0 = \{t\,x : 0 \leqslant t \leqslant 1\}$ and $C_1 = \{x : |x| \geqslant 1\}$. If Γ is the family of curves joining the boundary components of the ring $R(C_0, C_1)$, then

$$M(f(\Gamma)) \leqslant K\,M(\Gamma) = K\gamma_n(1/|x|),$$

by definition of K-quasiconformality and the definition of γ_n (using an inversion with respect to S^{n-1} to transform $R(C_0, C_1)$ onto $R_{G,n}(1/|x|)$ (Fig. 2.6).

Lemma 2.43 gives $M(f(\Gamma)) \geqslant \gamma_n(\frac{1}{|f(x)|})$. So,

$$\gamma_n(\frac{1}{|f(x)|}) \leqslant K\,\gamma_n\left(\frac{1}{|x|}\right),$$

i.e.,

$$\frac{1}{|f(x)|} \geqslant \gamma_n^{-1}\left(K\,\gamma_n\left(\frac{1}{|x|}\right)\right),$$

because γ_n is decreasing on $(1, +\infty)$. So,

$$|f(x)| \leqslant \frac{1}{\gamma_n^{-1}(K\,\gamma_n(\frac{1}{|x|}))}.$$

$\square$

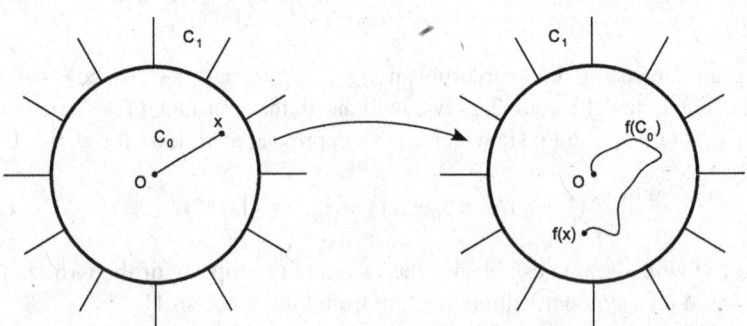

Fig. 2.6 Schwarz lemma

For $K = 1$ we have that $\varphi_{K,n}(r) = r$. We note that Theorem 2.44 was proved in [115, Theorem 3.1] for quasiregular mappings.

We will need a more general (Möbius invariant) form of this result. Below we will make use of the hyperbolic metric $\rho_{\mathbb{B}^n}$ of the unit ball and a variant of the Schwarz lemma formulated in terms of this metric. We have the formula [158, (2.18), 2.52]

$$\text{th}^2 \frac{1}{2}\rho(x, y) = \frac{|x - y|^2}{|x - y|^2 + (1 - |x|^2)(1 - |y|^2)} \tag{2.12}$$

for $x, y \in \mathbb{B}^n$.

Theorem 2.45 ([158, Theorem 11.2 (1)]) *If $f : D \to G$, $D, G \in \{\mathbb{B}^n, \mathbb{H}^n\}$ is K-quasiregular and $x, y \in D$, then*

$$\text{th} \frac{1}{2}\rho_G(f(x), f(y)) \leqslant \varphi_{K,n}(\text{th} \frac{1}{2}\rho_D(x, y)).$$

Proof In [158, Theorem 11.2 (1)] the result is formulated only in the case when $D = G = \mathbb{B}^n$. By Möbious invariance of the hyperbolic metric the same proof also holds in the present case. □

Now we are ready to prove the local Hölder continuity of a quasiconformal mapping.

Theorem 2.46 ([115]) *Suppose that f is a bounded and quasiconformal mapping in a domain $G \subseteq \mathbb{R}^n$ and that F is a compact subset of G. Set $\alpha = K_I(f)^{1/(1-n)}$ and $C = \lambda_n^{1-\alpha}d(F, \partial G)^{-\alpha}d(fG)$, where λ_n is the Grötzsch constant. Then f satisfies the Hölder condition*

$$|f(x) - f(y)| \leqslant C|x - y|^\alpha, \quad x \in F, \quad y \in G. \tag{2.13}$$

Proof Set $r = d(F, \partial G)$. The main case is $|x - y| < r$. Define $g : \mathbb{B}^n \to \mathbb{B}^n$ by

$$g(z) = \frac{f(x + rz) - f(x)}{d(fG)}, \quad |z| < 1.$$

Then $g(0) = 0$, $|g(z)| \leqslant 1$, and $K_I(g) \leqslant K_I(f)$.

Note that g is not necessarily onto $\mathbb{B}^n$. However, we can still apply the Schwarz lemma, Theorem 2.44, and use the estimate (2.8) to get

$$|g(z)| \leqslant \lambda_n^{1-\alpha}|z|^\alpha.$$

Set $z = \frac{y-x}{r} \in \mathbb{B}^n$. This gives

$$\frac{|f(y) - f(x)|}{d(fG)} \leqslant \lambda_n^{1-\alpha}\frac{|y - x|^\alpha}{r^\alpha}$$

i.e.,

$$|f(y) - f(x)| \leqslant \lambda_n^{1-\alpha} \cdot \frac{d(fG)}{d(F, \partial G)^\alpha} \cdot |y - x|^\alpha.$$

Now we consider the easier case $|x - y| \geqslant r$. We have

$$|f(x) - f(y)| \leqslant d(fG) \leqslant \frac{|x - y|^\alpha}{r^\alpha} d(fG) \leqslant \lambda_n^{1-\alpha} \cdot \frac{d(fG)}{d(F, \partial G)^\alpha} |x - y|^\alpha,$$

because $\lambda_n \geqslant 1$ (see Lemma 2.18). □

Note 2.47 Theorem 2.46 is also true if we replace $K_I(f)$ by $K(f)$.

Theorem 2.48 ([49, Theorem 11] and [155, Theorem 18.1, p. 63]) *(The distortion theorem) For every $K \geqslant 1$ and $n \in \mathbb{N}$, $n \geqslant 2$, there exists a function $\theta_{K,n} : (0, 1) \to \mathbb{R}^1$ with the following properties:*

1. *$\theta_{K,n}$ is increasing,*
2. *$\lim_{r \to 0} \theta_{K,n}(r) = 0$,*
3. *$\lim_{r \to 1} \theta_{K,n}(r) = \infty$.*
4. *Let D and D' be proper subdomains of $\mathbb{R}^n$ and let $f : D \to D'$ be K-quasiconformal. If $x, y \in D$ such that $0 < |y - x| < d(x, \partial D)$, then*

$$\frac{|f(y) - f(x)|}{d(f(x), \partial D')} \leq \theta_{K,n}\left(\frac{|y - x|}{d(x, \partial D)}\right),$$

$$\frac{|f(y) - f(x)|}{d(f(y), \partial D')} \leq \theta_{K,n}\left(\frac{|y - x|}{d(x, \partial D)}\right).$$

Moreover, we can choose

(A)

$$\theta_{K,n}(r) = \frac{1}{\tau_n^{-1}\left(K \, \omega_{n-1}\left(\log \frac{1}{r}\right)^{1-n}\right)}, \qquad where \; 0 < r < 1$$

and also

(B)

$$\theta_{K,n}(r) = \frac{1}{\psi^{-1}\left(\Phi\left(\frac{1}{r}\right)^\alpha\right)} \qquad for \; 0 < r < 1.$$

We give two proofs of this result, based on the same idea, leading to two different functions $\theta_{K,n}$. Our motivation for giving two proofs is coming from the fact that the function obtained in the second proof gives a better estimate.

Proof (A) Suppose that $f : D \to D'$ and x, y are as in (4). We abbreviate $d = d(x, \partial D)$, $d' = d(f(x), \partial D')$, and $d'' = d(f(y), \partial D')$. Let A be the spherical ring $\{z \mid |y - x| < |z - x| < d\}$. Then $A \subset D$. Setting $C_0 = f B^n(x, |y - x|)$ and $C_1 = D' \setminus f B^n(x, d)$, we have $f A = R(C_0, C_1)$. Here C_0 contains $f(x)$ and $f(y)$ while C_1 contains ∞ and points $b', b'' \in \partial D'$ such that $|f(x) - b'| = d'$, $|f(y) - b''| = d''$.

By 2.21

$$M(\Gamma_{fA}) \geqslant \tau_n \left(\frac{d'}{|f(x) - f(y)|} \right), \tag{2.14}$$

$$M(\Gamma_{fA}) \geqslant \tau_n \left(\frac{d''}{|f(x) - f(y)|} \right). \tag{2.15}$$

But, K-quasiconformality of f gives $M(\Gamma_{fA}) \leqslant K \cdot M(\Gamma_A)$ (Fig. 2.7).

Since $M(\Gamma_A) = \omega_{n-1} \left(\log \frac{d}{|y-x|} \right)^{1-n}$,

$$K \omega_{n-1} \left(\log \frac{d}{|y - x|} \right)^{1-n} \geqslant \tau_n \left(\frac{d'}{|f(x) - f(y)|} \right).$$

Since τ_n is a homeomorphism, we have

$$\tau_n^{-1} \left(K \omega_{n-1} \left(\log \frac{d}{|y - x|} \right)^{1-n} \right) \leqslant \frac{d'}{|f(x) - f(y)|},$$

i.e.,

$$\frac{|f(x) - f(y)|}{d'} \leqslant \frac{1}{\tau_n^{-1} \left(K \omega_{n-1} \left(\log \frac{d}{|y-x|} \right)^{1-n} \right)}.$$

Defining

$$\theta_{K,n}(r) = \frac{1}{\tau_n^{-1} \left(K \omega_{n-1} \left(\log \frac{1}{r} \right)^{1-n} \right)}, \quad \text{where } 0 < r < 1$$

we obtain the first inequality. The other inequality can be proved analogously, using (2.15) instead of (2.14).

Properties (1)–(3) follow from the corresponding properties of τ_n in 2.20. $\square$

Proof ((B) [33, p. 248]) Now we use, instead of a spherical ring A, a bounded Grötzsch ring $R = R_G(C_0, C_1)$, where $C_1 = S^{n-1}(x, a)$, where $|y - x| < a < d = d(x, \partial D)$ and C_0 is a line segment from x to y. In this case,

Fig. 2.7 The distortion
theorem

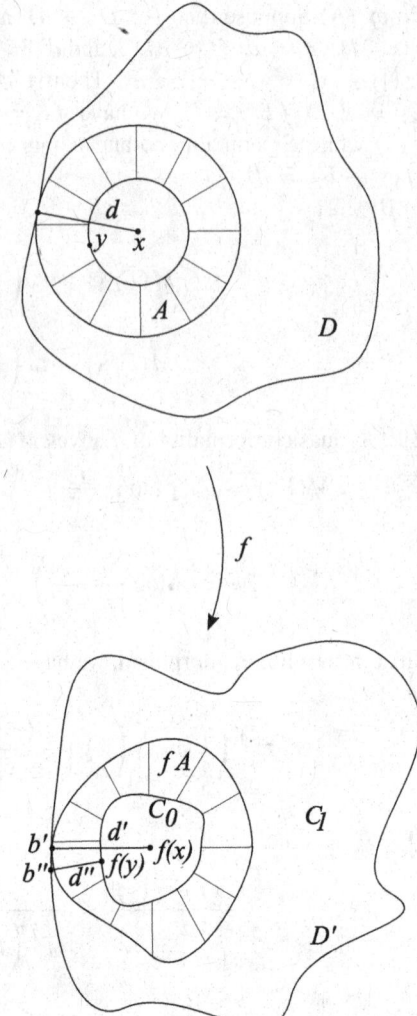

$$mod(R) = \log \Phi \left(\frac{a}{|x - y|} \right).$$

Now, $f(R) = R'$ is a ring with components C_0' and C_1' such that C_0' contains $f(x)$
and $f(y)$ and C_1' contains a point whose distance from $f(x)$ is smaller than d'.

Let $\alpha = K^{1/(1-n)}$. By the extremal property of Teichmüller ring (see Theorem 2.21) we have

$$mod(R') \leqslant \log \Psi \left(\frac{d'}{|f(x) - f(y)|} \right).$$

Since $mod(R') \geqslant \alpha \, mod(R)$, we obtain

$$\Psi\left(\frac{d'}{|f(x) - f(y)|}\right) \geqslant \Phi\left(\frac{a}{|x - y|}\right)^\alpha,$$

hence

$$\frac{d'}{|f(x) - f(y)|} \geqslant \Psi^{-1}\left(\Phi\left(\frac{a}{|x - y|}\right)^\alpha\right),$$

i.e.,

$$\frac{|f(x) - f(y)|}{d'} \leqslant \frac{1}{\Psi^{-1}\left(\Phi\left(\frac{a}{|x-y|}\right)^\alpha\right)},$$

letting $a \to d$, we get

$$\frac{|f(x) - f(y)|}{d'} \leqslant \frac{1}{\Psi^{-1}\left(\Phi\left(\frac{d}{|x-y|}\right)^\alpha\right)}$$

and we obtained the desired estimate with

$$\theta_{K,n}(r) = \frac{1}{\Psi^{-1}\left(\Phi\left(\frac{1}{r}\right)^\alpha\right)} \quad \text{for } 0 < r < 1.$$

$\square$

We reserve the notation $\theta_{K,n}$ for the function obtained in proof B.

We see that

$$\lim_{r \to 0} \frac{\theta_{K,n}(r)}{r^\alpha} = \lim_{r \to 0} \left(\frac{\Phi\left(\frac{1}{r}\right)^\alpha}{\Psi^{-1}\left(\Phi\left(\frac{1}{r}\right)^\alpha\right)} \cdot \frac{\left(\frac{1}{r}\right)^\alpha}{\Phi\left(\frac{1}{r}\right)^\alpha}\right) =$$

$$\lim_{t \to +\infty} \left(\frac{t}{\Phi(t)}\right)^\alpha \cdot \lim_{s \to +\infty} \frac{\Psi(s)}{s},$$

but both limits are finite (Lemma 2.17), so

$$\theta_{K,n}(r) = O(r^\alpha). \tag{2.16}$$

Hence, we derived behavior of $\theta_{K,n}(r)$ for small r.

We can find another expression for $\theta_{K,n}$, in terms of γ, τ. Namely,

$$\theta_{K,n}(t) = \frac{1}{\tau^{-1}\left(K \gamma\left(\frac{1}{t}\right)\right)}, \quad \text{for } 0 < t < 1.$$

This follows from the equation

$$\Psi^{-1}\left(\Phi\left(\frac{1}{r}\right)^{\alpha}\right) = \tau^{-1}\left(K\gamma\left(\frac{1}{r}\right)\right)$$

and by the definition of Ψ, we have

$$\Psi(u) = \exp\left(\frac{\tau(u)}{\omega_{n-1}}\right)^{1/(1-n)}$$

and now we obtain

$$\Phi\left(\frac{1}{r}\right)^{\alpha} = \exp\left(\frac{K\gamma\left(\frac{1}{r}\right)}{\omega_{n-1}}\right)^{1/(1-n)},$$

i.e.,

$$\alpha\log\Phi\left(\frac{1}{r}\right) = \left(\frac{K\gamma\left(\frac{1}{r}\right)}{\omega_{n-1}}\right)^{1/(1-n)},$$

which is equivalent to

$$\alpha\left(\frac{\gamma\left(\frac{1}{r}\right)}{\omega_{n-1}}\right)^{1/(1-n)} = K^{1/(1-n)}\left(\frac{\gamma\left(\frac{1}{r}\right)}{\omega_{n-1}}\right)^{1/(1-n)}$$

by the definition of Φ.

Next, we relate the function $\varphi_{T,n}$ to $\theta_{K,n}$:

$$\theta_{K,n}(t) = \frac{\varphi_{T,n}(t)^2}{1 - \varphi_{T,n}(t)^2}, \quad 0 < t < 1, \ T = 2^{n-1}K.$$

This follows easily from the following equation (see [10, 8.70 (7) (a)])

$$\tau_n^{-1}\left(K\gamma_n\left(\frac{1}{r}\right)\right) = \frac{\psi_{1/T,n}(r')^2}{\varphi_{T,n}(r)^2},$$

where $r' = \sqrt{1-r^2}$ and $\psi_{K,n}(r) = \sqrt{1 - \varphi_{1/K,n}(r')^2}$. $\square$

2.4   Moduli of Continuity in the Plane

This section considers a natural topic, the relationship between moduli of continuity of mappings that are continuous on the boundary of the ball and quasiregular inside the ball and it is based on the paper [88]. This was also considered by Heinonen et al. in [64, Theorem 14.45, p.273]. In [88], it was proven that $|f|^p$ is subharmonic whenever f is QR, and moreover the optimal exponent $p < 1$ was explicitly determined.

It is well known that if f is a complex-valued harmonic function defined in a domain G of the complex plane $\mathbb{C}$, then $|f|^p$ is subharmonic for $p \geq 1$, and that in the general case it is not subharmonic for $p < 1$. However, if f is holomorphic, then $|f|^p$ is subharmonic for every $p > 0$ (see [40, Example 1 and Example 2, pp. 7–8]).

Here we consider quasiregular harmonic functions in the plane.

We shall follow the practice of the planar case to define k-quasiregular mapping for $0 < k < 1$ as follows: A mapping $f : G \longrightarrow \mathbb{C}$ is k-quasiregular if it is absolutely continuous on lines in G and

$$|\bar{\partial} f(z)| \leq k|\partial f(z)|, \quad \text{for almost all } z \in G, \tag{2.17}$$

where

$$\bar{\partial} f(z) = \frac{1}{2}\left(\frac{\partial f}{\partial x} + i\frac{\partial f}{\partial y}\right) \quad \text{and} \quad \partial f(z) = \frac{1}{2}\left(\frac{\partial f}{\partial x} - i\frac{\partial f}{\partial y}\right), \quad z = x + iy.$$

As shown in [94], k-quasiregularity in this sense is the same thing as K-quasiregularity with $K = (1 + k)/(1 - k)$ in the sense of Definition 2.25.

For a continuous function $f : \overline{\mathbb{D}} \to \mathbb{C}$ harmonic in $\mathbb{D}$, we define two moduli of continuity:

$$\omega(f, \delta) = \sup\{|f(e^{i\theta}) - f(e^{it})| : |e^{i\theta} - e^{it}| \leq \delta, \ t, \theta \in \mathbb{R}\}, \quad \delta \geq 0,$$

and

$$\tilde{\omega}(f, \delta) = \sup\{|f(z) - f(w)| : |z - w| \leq \delta, \ z, w \in \overline{\mathbb{D}}\}, \quad \delta \geq 0.$$

We prove that $|f|^p$ is subharmonic for $p \geq 4k/(1 + k)^2 =: q$ as well as that the exponent $q\ (< 1)$ is the best possible (see Theorem 2.53). The fact that $q < 1$ enables us to prove that if f is quasiregular in the unit disk $\mathbb{D}$ and continuous on $\overline{\mathbb{D}}$, then $\tilde{\omega}(f, \delta) \leq C\,\omega(f, \delta)$ for some constant C; see Theorem 2.60.

If $u : G \longrightarrow \mathbb{R}$, $u > 0$ is a real valued function of the class C^2 defined on a domain G in $\mathbb{C}$, then

$$|\nabla u|^2 = \left(\frac{\partial u}{\partial x}\right)^2 + \left(\frac{\partial u}{\partial y}\right)^2 = \left|\frac{\partial u}{\partial x} - i\frac{\partial u}{\partial y}\right|^2 = 4|\partial u|^2,$$

$$|\nabla u|^2 = \left(\frac{\partial u}{\partial x}\right)^2 + \left(\frac{\partial u}{\partial y}\right)^2 = \left|\frac{\partial u}{\partial x} + i\frac{\partial u}{\partial y}\right|^2 = 4|\bar{\partial} u|^2,$$

$$\Delta = \left(\frac{\partial}{\partial x} - i\frac{\partial}{\partial y}\right)\left(\frac{\partial}{\partial x} + i\frac{\partial}{\partial y}\right) = 4\frac{\partial^2}{\partial z \partial \bar{z}},$$

or in the more compact form

$$|\nabla u|^2 = 4|\partial u|^2 = 4|\bar{\partial} u|^2 \qquad (2.18)$$

and

$$\Delta u = 4\partial\bar{\partial} u. \qquad (2.19)$$

Lemma 2.49 *If G is a domain in $\mathbb{C}$ and if $u : G \longrightarrow \mathbb{R}$ is a function of class C^2 such that $u > 0$, then for every $\alpha \in \mathbb{R}$,*

$$\Delta(u^\alpha) = \alpha u^{\alpha-1}\Delta u + \alpha(\alpha - 1)u^{\alpha-2}|\nabla u|^2, \qquad (2.20)$$

Proof

$$\Delta(u^\alpha) = \sum_i (u^\alpha)_{x_i x_i} = \sum_i ((u^\alpha)_{x_i})_{x_i} = \sum_i (\alpha\, u^{\alpha-1} u_{x_i})_{x_i}$$
$$= \alpha \sum_i ((\alpha - 1)u^{\alpha-2}(u_{x_i})^2 + u^{\alpha-1}u_{x_i x_i})$$
$$= \alpha u^{\alpha-1}\Delta u + \alpha(\alpha - 1)u^{\alpha-2}|\nabla u|^2.$$

$\square$

Lemma 2.50 *If $f = g + \bar{h}$, where g and h are complex valued holomorphic functions defined on domain G in $\mathbb{C}$, then*

$$\Delta(|f|^2) = 4(|g'|^2 + |h'|^2).$$

Proof Since $|f|^2 = (g + \bar{h})(\bar{g} + h)$, we have

$$\Delta(|f|^2) = 4\partial(\overline{h'}(\bar{g} + h) + (g + \bar{h})\overline{g'})$$
$$= 4(\overline{h'}h + g\overline{g'})$$
$$= 4(|g'|^2 + |h'|^2).$$

$\square$

Lemma 2.51 *If $f = g + \bar{h}$, where g and h are complex valued holomorphic functions defined on domain G in $\mathbb{C}$, then*

$$|\nabla(|f|^2)|^2 = 4(|g'|^2 + |h'|^2)|f|^2 + 8\operatorname{Re}(\overline{g'}h'f^2).$$

Proof We have

$$
\begin{aligned}
|\nabla(|f|^2)|^2 &= 4|\partial(|f|^2)|^2 \\
&= 4|\partial((g+\bar{h})(\bar{g}+h))|^2 \\
&= 4|g'\bar{f}+fh'|^2 \\
&= 4(|g'|^2+|h'|^2)|f|^2+8\,\mathrm{Re}\,(\overline{g'}h'f^2).
\end{aligned}
$$

$\square$

Lemma 2.52 *If $f = g + \bar{h}$, where g and h are complex valued holomorphic functions defined in domain G in $\mathbb{C}$, then*

$$
\Delta(|f|^p) = p^2(|g'|^2+|h'|^2)|f|^{p-2}+2p(p-2)|f|^{p-4}\mathrm{Re}\,(\overline{g'}h'f^2),
$$

whenever $f \neq 0$.

Proof We take $\alpha = p/2$, $u = |f|^2$ and then use (2.20), Lemma 2.50, and Lemma 2.51 to get the result. $\square$

Theorem 2.53 ([88]) *If f is a complex-valued k-quasiregular harmonic function defined on a domain $G \subset \mathbb{C}$, and $q = 4k/(k+1)^2$, then $|f|^q$ is subharmonic. The exponent q is optimal.*

Proof If $u(z_0) = |f(z_0)|^2 = 0$, then (*) from the Preliminaries holds. If $u(z_0) > 0$, then there exists a neighborhood U of z_0 such that u is of class $C^2(U)$ (because the zeroes of u are isolated), and then we can prove that $\Delta u \geq 0$ on U.

Thus the proof reduces to proving that $\Delta u(z) \geq 0$ whenever $u(z) > 0$.

We have to prove that $\Delta(|f|^p) \geqslant 0$, where $p = 4k/(1+k)^2$. Since $p - 2 < 0$, we get from Lemma 2.52 that

$$
\begin{aligned}
\Delta(|f|^p) &\geqslant p^2(|g'|^2+|h'|^2)|f|^{p-2}+2p(p-2)|f|^{p-4}|g'|\cdot|h'|\cdot|f|^2 \\
&= p^2|g'|^2(m^2+1)|f|^{p-1}+2p(p-2)|g'|^2|f|^{p-2}m \\
&= p|g'|^2|f|^{p-2}[p(1+m)+2(p-2)m],
\end{aligned}
$$

where $m = |h'|/|g'| \leqslant k$. The function $m \mapsto p(1+m^2)+2(p-2)m$ has a negative derivative (because $p < 1$ and $m < 1$), and this implies that

$$
(1+m^2)p+2(p-2)m \geqslant (1+k^2)p+2(p-2)k.
$$

Observe now that $(1+k^2)p+2(p-2)k \geqslant 0$ if and only if $p \geqslant 4k/(1+k)^2$. This proves that $|f|^q$ is subharmonic. To prove that q is optimal consider $f(z) = z + k\bar{z}$. Note that

$$
|f(1)| = |1+k\cdot\bar{1}| = 1+k.
$$

By Lemma 2.52 for $g(z) = z$ and $h(z) = kz$, we get

$$\Delta(|f|^p)(1) = p^2(1 + k^2)(1 + k)^{p-2} + 2p(p - 2)(1 + k)^{p-2}k.$$

It follows that $\Delta(|f|^p)(1) \geqslant 0$ if and only if

$$p(1 + k^2) + 2(p - 2)k \geqslant 0,$$

which, as indicated above, is equivalent to $p \geqslant q$.   □

Lemma 2.54 (Jordan Inequality) *For $0 < x \leq \pi/2$ there holds*

$$\frac{2}{\pi} \leq \frac{\sin(x)}{x} < 1.$$

Lemma 2.55 *For $t_1, t_2 \in \mathbb{R}$ such that $|t_1 - t_2| \leq \pi$ there holds*

$$|t_1 - t_2| \leq \frac{\pi}{2}|e^{it_1} - e^{it_2}|,$$

$$|e^{it_1} - e^{it_2}| \leq |t_1 - t_2|.$$

Proof For $t_1 = t_2$ the conclusion of the lemma holds. Assume that $t_1 \neq t_2$. Then, we have

$$|e^{it_1} - e^{it_2}| = 2\sin(|t_1 - t_2|/2)$$

$$|e^{it_1} - e^{it_2}|\frac{\sin(|t_1 - t_2|/2)}{|t_1 - t_2|/2}|t_1 - t_2| \leq |t_1 - t_2|,$$

$$|t_1 - t_2| = \frac{|e^{it_1} - e^{it_2}|}{\left(\frac{|e^{it_1} - e^{it_2}|}{|t_1 - t_2|}\right)} = \frac{|e^{it_1} - e^{it_2}|}{\left(\frac{\sin(|t_1 - t_2|/2)}{|t_1 - t_2|/2}\right)} \leq |e^{it_1} - e^{it_2}|\frac{\pi}{2}.$$

□

Lemma 2.56 *For all $\delta_1, \delta_2 \geq 0$, we have*

$$\omega_0(f, \delta_1 + \delta_2) \leq \omega_0(f, \delta_1) + \omega_0(f, \delta_2).$$

Proof Let $\delta_1, \delta_2 > 0$. Let us define the sets

$$S(\delta) = \{|f(e^{it_1}) - f(e^{it_2})| : |t_1 - t_2| \leq \delta\}.$$

Choose a $a \in S(\delta_1 + \delta_2)$. Then there are $t_1, t_2 \in \mathbb{R}$ such that $|t_1 - t_2| \leq \delta_1 + \delta_2$ and $|f(e^{it_1}) - f(e^{it_2})| = a$. There is a $t \in \mathbb{R}$ such that $|t_1 - t| \leq \delta_1$ and $|t - t_2| \leq \delta_2$. For

$$b = |f(e^{it_1}) - f(e^{it})|, \quad c = |f(e^{it}) - f(e^{it_2})|,$$

we have that $b \in S(\delta_1)$ and $c \in S(\delta_2)$. Therefore,

$$b \leq \sup S(\delta_1) = \omega_0(f, \delta_1),$$

$$c \leq \sup S(\delta_2) = \omega_0(f, \delta_2),$$

$$a \leq b + c \leq \omega_0(f, \delta_1) + \omega_0(f, \delta_2).$$

Because all elements of $S(\delta_1 + \delta_2)$ are less or equal to $\omega_0(f, \delta_1) + \omega_0(f, \delta_2)$, we have that

$$\omega_0(f, \delta_1 + \delta_2) = \sup S(\delta_1 + \delta_2) \leq \omega_0(f, \delta_1) + \omega_0(f, \delta_2).$$

$\square$

Corollary 2.57 *For every $\delta \geq 0$, we have the following inequalities*

1. $\omega_0(f, m\delta) \leq m\omega(f, \delta)$ *for every $m \in \mathbb{N}$.*
2. $\omega_0(f, \lambda\delta) \leq n\,\omega_0(f, \delta)$ *for every $\lambda \in \mathbb{R}$ and $n \in \mathbb{N}$ such that $0 < \lambda \leq n$.*
3. $\omega_0(f, \lambda\delta) \leq (\lambda + 1)\omega_0(f, \delta)$ $\lambda \in \mathbb{R}$ *such that $\lambda > 0$.*

Lemma 2.58 *We have the following inequalities:*

$$\omega_0(f, \delta) \leq \omega(f, \delta) \leq 2\omega_0(f, \delta).$$

Proof Let

$$S(\delta) = \{|f(e^{it_1}) - f(e^{it_2})| \,:\, |e^{it_1} - e^{it_2}| \leq \delta\},$$

$$S_0(\delta) = \{|f(e^{it_1}) - f(e^{it_2})| \,:\, |t_1 - t_2| \leq \delta\}.$$

If $a \in S$, then there are $t_1, t_2 \in \mathbb{R}$ such that

$$|e^{it_1} - e^{it_2}|,$$

$$a = |f(e^{it_1}) - f(e^{it_2})|.$$

Hence,

$$|t_1 - t_2| \leq \frac{\pi}{2}|e^{it_1} - e^{it_2}| \leq \frac{\pi}{2}\delta$$

and therefore $a \in S_0(\delta\pi/2)$. This means that $S(\delta) \subseteq S_0(\delta\pi/2)$ and for this reason

$$\omega(f, \delta) = \sup S(\delta) \leq \sup S_0(\delta\pi/2) = \omega_0(f, \delta\pi/2).$$

If $b \in S_0$, then there are $\tau_1, \tau_2 \in \mathbb{R}$ such that $|\tau_1 - \tau_2| \leq \delta$ and

$$b = |f(e^{i\tau_1}) - f(e^{i\tau_2})|.$$

Note that

$$|e^{i\tau_1} - e^{i\tau_2}| \leq |\tau_1 - \tau_2| \leq \delta$$

and therefore $b \in S(\delta)$. This means that $S_0(\delta) \subseteq S(\delta)$ and

$$\omega_0(f, \delta) = \sup S_0(\delta) \leq \sup S(\delta) = \omega(f, \delta).$$

Finally,

$$\omega_0(f, \delta) \leq \omega(f, \delta) \leq \omega_0(f, \delta\pi/2) \leq 2\omega_0(\delta).$$

$\square$

From these lemmas it follows that

$$\omega(f, \lambda\delta) \leq 2\lambda\omega(f, \delta), \quad \lambda \geq 1, \delta \geq 0, \tag{2.21}$$

and

$$\omega(f, \delta_1 + \delta_2) \leq 2\omega(f, \delta_1) + 2\omega(f, \delta_2), \quad \delta_1, \delta_2 \geq 0. \tag{2.22}$$

As a consequence of (2.21) we have, for $0 < p < 1$,

$$\int_x^\infty \frac{\omega(f, t)^p}{t^2} \, dt \leq C \frac{\omega(f, x)^p}{x}, \quad x > 0, \tag{2.23}$$

where C depends only on p.

For a continuous function $f : \overline{\mathbb{D}} \longrightarrow \mathbb{C}$ harmonic in $\mathbb{D}$ we already defined moduli of continuity $\omega(f, \delta)$ and $\tilde{\omega}(f, \delta)$.

Clearly $\omega(f, \delta) \leq \tilde{\omega}(f, \delta)$, but the reverse inequality need not hold. To see this consider the function

$$f(re^{i\theta}) = \sum_{n=1}^\infty \frac{(-1)^n r^n \cos n\theta}{n^2}, \quad re^{i\theta} \in \mathbb{D}. \tag{2.24}$$

Recall that the polar form of the Laplacian is

$$\Delta f(r\, e^{i\theta}) = \frac{\partial^2 f}{\partial r^2} + \frac{1}{r} \frac{\partial f}{\partial r} + \frac{1}{r^2} \frac{\partial^2 f}{\partial \theta^2}.$$

The series from (2.24) can be differentiated by variables r and θ item by item arbitrary many times inside disk $r < 1$. Any summand of this series is harmonic and therefore f is harmonic in the disk $\mathbb{D}$. The same series is uniformly convergent in the disk $r \leqq 1$ and all summands are continuous in this disk. Therefore, f is continuous in $\overline{\mathbb{D}}$.

The function $v(\theta) = f(e^{i\theta})$, $|\theta| < \pi$, is differentiable, and

$$\frac{dv}{d\theta} = \sum_{n=1}^{\infty} \frac{(-1)^{n-1} \sin n\theta}{n}$$
$$= \frac{\theta}{2}, \qquad |\theta| < \pi.$$

The first equality follows from Abel's test because $1/n$ (uniformly) and monotonously tends to zero and $(-1)^{n-1} \sin(n\theta)$ is uniformly bounded. The second equality is the expansion of the function $\theta/2$ into Fourier series. Namely, using the fact that the function $\theta/2$ is odd we can conclude that

$$a_0 = \frac{1}{2\pi} \int_{-\pi}^{\pi} \frac{\theta}{2} \, d\theta = 0,$$

and for $n \geq 1$

$$a_n = \frac{1}{\pi} \int_{-\pi}^{\pi} \frac{\theta}{2} \cos(n\theta) \, d\theta = 0,$$

$$b_n = \frac{1}{\pi} \int_{-\pi}^{\pi} \frac{\theta}{2} \sin(n\theta) \, d\theta = \frac{1}{\pi} \int_{0}^{\pi} \theta \sin(n\theta) \, d\theta = \frac{(-1)^{n-1}}{n}.$$

The last equality is obtained by partial integration. Therefore,

$$\frac{\theta}{2} = a_0 + \sum_{n=1}^{\infty}(a_n \cos(n\theta) + b_n \sin(n\theta)) = \sum_{n=1}^{\infty} \frac{(-1)^{n-1} \sin(n\theta)}{n}, \qquad |\theta| < \pi.$$

From $v'(\theta) = \theta/2$ for all $\theta \in (-\pi, \pi)$ it follows that $|v'(c)| < \pi/2$ for all $c \in (-\pi, \pi)$. By Lagrange's theorem we have

$$|f(e^{i\theta}) - f(e^{it})| = |v(\theta) - v(t)| = |v'(c)(\theta - t)| \leq (\pi/2)|\theta - t|, \quad -\pi < \theta, t < \pi.$$

If $|e^{i\theta} - e^{it}| \leq \delta$, then we can choose $x, y \in \mathbb{R}$ such that $e^{i\theta} = e^{ix}$, $e^{it} = e^{iy}$ and $|x - y| \leq \pi$. For this reason and by Lemma 2.55,

$$|f(e^{i\theta}) - f(e^{it})| = |f(e^{ix}) - f(e^{iy})| \leq \frac{\pi}{2}|x - y| \leq \frac{\pi^2}{4}|e^{ix} - e^{iy}| = \frac{\pi^2}{4}|e^{i\theta} - e^{it}|,$$

and therefore $\omega(f, \delta) \leq \frac{\pi^2}{4}\delta$, $\delta > 0$.

On the other hand, the inequality $\tilde{\omega}(f,\delta) \leq CM\delta$, $C = $ const, does not hold because it implies that $|\partial f/\partial r| \leq CM$, which is not true (when $r \to 1-$) because

$$\frac{\partial}{\partial r} f(re^{i\theta}) = \sum_{n=1}^{\infty} \frac{r^{n-1}}{n} = -\frac{\ln(1-r)}{r}, \quad \text{for } \theta = \pi,\ 0 < r < 1.$$

The second equality is the direct consequence of the Taylor's expansion

$$-\ln(1-r) = \sum_{n=1}^{\infty} \frac{r^n}{n}, \quad -1 \leq r < 1.$$

However, as was proved by Rubel, Shields, and Taylor [137] as well as Tamrazov [151], that if f is a holomorphic function, then $\tilde{\omega}(f,\delta) \leq C\omega(f,\delta)$, where C is independent of f and δ.

The next theorem will be deduced from Theorem 2.53 by use of some simple properties of the modulus $\omega(f,\delta)$. Let

$$\omega_0(f,\delta) = \sup\{|f(e^{i\theta}) - f(e^{it})| : |\theta - t| \leq \delta,\ t, \theta \in \mathbb{R}\}.$$

To extend this result to HQR functions we need the following consequence of the harmonic Schwarz lemma (see [23, p. 125]).

Lemma 2.59 *If $h : \mathbb{D} \longrightarrow \mathbb{R}$ is a function harmonic and bounded in the unit disk, with $h(0) = 0$, then $|h(\xi)| \leqslant (4/\pi)||h||_\infty$, for $\xi \in \mathbb{D}$.*

Theorem 2.60 ([88]) *Let $f : \mathbb{D} \longrightarrow \mathbb{C}$ be a k-quasiregular harmonic function which has a continuous extension on $\overline{\mathbb{D}}$, then there is a constant C depending only on k such that $\tilde{\omega}(f,\delta) \leq C\omega(f,\delta)$.*

Proof It is enough to prove that $|f(z) - f(w)| \leqslant C\omega(f, |z - w|)$ for all $z, w \in \overline{\mathbb{D}}$, where C depends only on k. Assume first that $z = r \in (0,1)$ and $|w| = 1$. Then, by Theorem 2.53, the function $\varphi(\xi) = |f(w) - f(\xi)|^q$, where $q = 4k/(1+k)^2 < 1$ is subharmonic in $\mathbb{D}$ and continuous on $\overline{\mathbb{D}}$, so that

$$\varphi(\zeta) \leqslant \frac{1}{2\pi} \int_{\partial \mathbb{D}} \frac{(1-r^2)\varphi(\zeta)}{|\zeta - r|^2} |d\zeta|.$$

For all $a, b > 0$ holds

$$\left(\frac{a}{a+b}\right)^q + \left(\frac{b}{a+b}\right)^q \geq \frac{a}{a+b} + \frac{b}{a+b} = 1,$$

and therefore

$$\left(\frac{a}{a+b}\right)^q + \left(\frac{b}{a+b}\right)^q \geq 1$$

and

$$(a+b)^q \le a^q + b^q.$$

The last inequality also holds if at least one of a, b is zero. Therefore, it holds for $a, b \ge 0$. Since, by (2.22),

$$\varphi(\zeta) \le (\omega(f, |w - r| + |r - \zeta|))^q$$
$$\le 2^q \omega(f, |w - r|)^q + 2^q \omega(f, |r - \zeta|)^q,$$

we have

$$\varphi(z) \le 2^q \omega(f, |w - r|)^q + \frac{2^{q-1}}{\pi} \int_{\partial \mathbb{D}} \frac{(1 - r^2)\omega(f, |r - \zeta|)^q}{|\zeta - r|^2} |d\zeta|$$
$$\le 2^q \omega(f, |w - r|)^q + \frac{2^{q-1}}{\pi} \int_{-\pi}^{\pi} \frac{(1 - r^2)\omega(|r - e^{it}|)^q}{|e^{it} - r|^2} dt.$$

We have that

$$|r - e^{it}| = \sqrt{(1 - r)^2 + 4r \sin^2(t/2)} \asymp 1 - r + |t| \quad (0 < r < 1, |t| < \pi).$$

From this, (*) from the Preliminaries and (2.23) it follows that

$$\int_{-\pi}^{\pi} \frac{(1 - r^2)\omega(f, |r - e^{it}|)^q}{|e^{it} - r|^2} dt \le C_1 \int_0^{\pi} \frac{(1 - r)\omega(f, 1 - r + t)^q}{(1 - r + t)^2} dt$$
$$= C_1 \left(\int_0^{1-r} + \int_{1-r}^{\pi} \right) \frac{(1 - r^2)\omega(f, |r - e^{it}|)^q}{|e^{it} - r|^2} dt$$
$$\le C_2(\omega(1 - r))^q + C_2(1 - r) \int_{1-r}^{\infty} \frac{\omega(f, t)^q}{t^2} dt$$
$$\le C_3(\omega(f, 1 - r))^q$$
$$\le C_4(\omega(f, |w - z|))^q.$$

Thus $|f(w) - f(z)| \le C_5 \omega(f, |w - z|)$ provided $w \in \partial \mathbb{D}$ and $z \in (0, 1)$. Using a rotation and the continuity of f, we can extend this inequality to the case where $w \in \partial \mathbb{D}$ and $z \in \mathbb{D}$.

If $0 < |w| < 1$, we look at the function $h(\xi) - f(\xi w / |w|) - f(\xi z / |w|)$ for $|\xi| \le 1$. Note that this function is harmonic in $\mathbb{D}$, continuous on $\bar{\mathbb{D}}$, and $h(0) = 0$. Therefore, by the harmonic Schwarz lemma, the inequality (*) from the Preliminaries, and the previous case, we have

$$|f(w) - f(z)| = |h(w)|$$
$$\le (4/\pi)|w| \, \|h\|_\infty$$
$$\le C_6 |w| \omega(f, |w/|w| - z/|w||),$$
$$\le C_7 \omega(f, |w/|w| - z/|w||),$$
$$= C_7 \omega(f, |w - z|).$$

$\square$

2.5 Moduli of Continuity in Higher Dimensions

In [110], Olli Martio gave sufficient conditions for K-quasiconformality of a harmonic mapping from the unit disk onto itself. In the same paper he posed a question answered fully in the paper [14] where the following result is proved.

Theorem 2.61 *If u is harmonic in the unit disk $\mathbb{D}$, and f is the restriction of u on $S^1 = \partial\mathbb{D}$, find necessary and sufficient conditions that $\lim_{z\to\zeta} u_r(z)$ and $\lim_{z\to\zeta} u_\theta(z)$ exist at each boundary point $\zeta \in S^1$.*

Here, u_r and u_θ are derivatives of u with respect to r and θ.

Moreover, Martio's paper [110] served as a motivation for further study of such phenomena. The following general, but natural question has been considered in several papers ([12–15]):

Can one control the modulus of continuity ω_u of a harmonic quasiregular mapping u in $\mathbb{B}^n$, which is continuous on $\overline{\mathbb{B}^n}$, by the modulus of continuity ω_f of its restriction to the boundary $\mathbb{S}^{n-1}$, i.e., is it true that $\omega_u \leq C\omega_f$?

In fact this problem has been studied extensively for harmonic functions and mappings without assumption of quasiregularity. We recall some of the known results. For the unit ball the answer is positive in the case $\omega(\delta) = \delta^\alpha$, $0 < \alpha < 1$ (Hölder continuity) and negative in the case $\omega(\delta) = L\delta$ (Lipschitz continuity). In fact, for bounded plane domains the answer is always negative for Lipschitz continuity (see [7]). However, it is proved in [68] that for general plane domains one has "logarithmic loss of control": $\omega_u(\delta) \leq C\omega_f(\delta)\log(1/\delta)$ for $\delta > 0$.

For $n \geq 2$,

$$P[\phi](x) = \int_{S^{n-1}} P(x,\xi)\phi(\xi)d\sigma(\xi), \quad x \in \mathbb{B}^n$$

where $P(x,\xi) = \frac{1-|x|^2}{|x-\xi|^n}$ is the Poisson kernel for the unit ball, and $\phi : S^{n-1} \to \mathbb{R}^n$ is a continuous mapping.

It was shown in [13] that Lipschitz continuity is preserved by harmonic extension, if the extension is quasiregular.

Theorem 2.62 ([13]) *Assume $\phi : S^{n-1} \to \mathbb{R}^n$ satisfies a Lipschitz condition:*

$$|\phi(\xi) - \phi(\eta)| \leq L|\xi - \eta|, \quad \xi, \eta \in S^{n-1}$$

and assume $u = P[\phi] : \mathbb{B}^n \to \mathbb{R}^n$ is K-quasiregular. Then

$$|u(x) - u(y)| \leq C'|x - y|, \quad x, y \in \mathbb{B}^n$$

where C' depends on L, K, and n only.

Proof We choose $x_0 = r \xi_0 \in \mathbb{B}^n$, $r = |x|$, $\xi_0 \in S^{n-1}$. Let $T = T_{x_0} r S^{n-1}$ be the $(n-1)$-dimensional tangent plane at x_0 to the sphere $r S^{n-1}$. We want to prove that

$$\|D(u|_T)(x_0)\| \leq C(n)L. \tag{2.25}$$

Without loss of generality we can assume $\xi_0 = e_n$ and $x_0 = r e_n$ by using a simple calculation

$$\frac{\partial}{\partial x_j} P(x, \xi) = \frac{-2x_j}{|x - \xi|^n} - n(1 - |x|^2) \frac{x_j - \xi_j}{|x - \xi|^{n+2}}.$$

Hence, for $1 \leq j < n$ we have

$$\frac{\partial}{\partial x_j} P(x_0, \xi) = n(1 - |x_0|^2) \frac{\xi_j}{|x_0 - \xi|^{n+2}}.$$

It is important to note that this kernel is odd in ξ (with respect to reflection $(\xi_1, \ldots, \xi_j, \ldots, \xi_n) \mapsto (\xi_1, \ldots, -\xi_j, \ldots, \xi_n)$), a typical fact for kernels obtained by differentiation. This observation and differentiation under integral sign give, for any $1 \leq j < n$,

$$\frac{\partial u}{\partial x_j}(x_0) = n(1 - r^2) \int_{S^{n-1}} \frac{\xi_j}{|x_0 - \xi|^{n+2}} \phi(\xi)\, d\sigma(\xi)$$

$$= n(1 - r^2) \int_{S^{n-1}} \frac{\xi_j}{|x_0 - \xi|^{n+2}} (\phi(\xi) - \phi(\xi_0))\, d\sigma(\xi).$$

Using the elementary inequality $|\xi_j| \leq |\xi - \xi_0|$, $(1 \leq j < n, \xi \in S^{n-1})$ and Lipschitz continuity of ϕ we get

$$\left| \frac{\partial u}{\partial x_j}(x_0) \right| \leq Ln(1 - r^2) \int_{S^{n-1}} \frac{|\xi_j|\,|\xi - \xi_0|}{|x_0 - \xi|^{n+2}}\, d\sigma(\xi)$$

$$\leq Ln(1 - r^2) \int_{S^{n-1}} \frac{|\xi - \xi_0|^2}{|x_0 - \xi|^{n+2}}\, d\sigma(\xi).$$

In order to estimate the last integral, we split S^{n-1} into two subsets $E = \{\xi \in S^{n-1} : |\xi - \xi_0| \leq 1 - r\}$ and $F = \{\xi \in S^{n-1} : |\xi - \xi_0| > 1 - r\}$. Since $|\xi - x_0| \geq 1 - |x_0|$ for all $\xi \in S^{n-1}$ we have

$$\int_E \frac{|\xi - \xi_0|^2}{|x_0 - \xi|^{n+2}}\, d\sigma(\xi) \leq (1 - r^2)^{-n-2} \int_E |\xi - \xi_0|^2 d\sigma(\xi)$$

$$\leq (1 - r^2)^{-n-2} \int_0^{1-r} \rho^2 \rho^{n-1}\, d\rho$$

$$\leq \frac{2}{n+1}(1 - r)^{-1}.$$

On the other hand, $|\xi - \xi_0| \leq C_n|\xi - x_0|$ for every $\xi \in F$, so

$$\int_F \frac{|\xi - \xi_0|^2}{|x_0 - \xi|^{n+2}} d\sigma(\xi) \leq C_n^{n+2} \int_F |\xi - \xi_0|^{-n} d\sigma(\xi)$$
$$\leq C_n' \int_{1-r}^2 \rho^{-n} \rho^{n-2} d\rho$$
$$\leq C_n'(1-r)^{-1}.$$

Combining these two estimates we get

$$\left| \frac{\partial u}{\partial x_j}(x_0) \right| \leq LC(n)$$

for $1 \leq j < n$. Due to rotational symmetry, the same estimate holds for every derivative in any tangential direction. This establishes estimate (2.25). Finally, K-quasiregularity gives

$$\|Du(x)\| \leq LKC(n).$$

Now the mean value theorem gives Lipschitz continuity of u.   □

In [14], the authors gave the estimate $L_u \leq KL_\phi$, where L_u, L_ϕ denote the Lipschitz constants of u, ϕ, respectively. Also in [14], Theorem 2.61 is extended to the n-dimensional case.

In [15], the following results were proved.

Theorem 2.63 ([15]) *There is a constant $q = q(K, n) \in (0, 1)$ such that $|u|^q$ is subharmonic in $\Omega \subset \mathbb{R}^n$ whenever $u : \Omega \to \mathbb{R}^n$ is a K-quasiregular harmonic map.*

Theorem 2.64 ([15]) *Let $u : \overline{\mathbb{B}}^n \to \mathbb{R}^n$ be a continuous map which is K-quasiregular and harmonic in $\mathbb{B}^n$. Then $\omega_u(\delta) \leq C\omega_f(\delta)$ for $\delta > 0$, where $f = u|_{\mathbb{S}^{n-1}}$ and C is a constant depending only on K, ω_f, and n.*

The proof of Theorem 2.63 is based on a linear algebra extremal problem.

It should be noted that every continuous map $u : \overline{\mathbb{B}}^n \to \overline{\Omega}$, which is HQC in $\mathbb{B}^n$ and whose domain Ω is bounded and has C^2 boundary, is Lipschitz continuous, see [76]. Moreover, every holomorphic quasiregular mapping on a domain $\Omega \subset \mathbb{C}^n$ ($n > 1$) with C^2 boundary is Lipschitz continuous, see [126]. This paper gives also an example of a holomorphic quasiregular map in a domain $\Omega \subset \mathbb{C}^2$ (with non-smooth boundary) which is not Lipschitz. It is worth pointing out that a global holomorphic quasiregular map is affine by a result of Marden and Rickman [105].

2.6 An Example of Non-Lipschitz HQC Mapping on the Unit Ball

Given the facts presented above, one is tempted to make the following conjecture: every HQC map $u : \overline{\mathbb{B}^n} \to \Omega$ is Lipschitz continuous. This is, however, false and an example for $n = 3$ is given in [15] on which we base this section.

Our next lemma is true for vectors in every real inner product space, although for our purposes here $M_3(\mathbb{R}) \cong \mathbb{R}^9$ is the only case of interest, where $M_3(\mathbb{R})$ is a set of all square matrices of order 3.

Recall that for $T \in M_3(\mathbb{R})$, $T \neq 0$, its normalization $\tilde{T}$ is defined by $\tilde{T} = \|T\|^{-1}T$.

Lemma 2.65 *Let $A \in M_3(\mathbb{R})$, $A \neq 0$, and $0 < \epsilon < \sqrt{2}$. Let $B_1, \ldots, B_k$ be matrices satisfying $\|\tilde{A} - \tilde{B}_j\| < \epsilon$, $1 \leq j \leq k$. Then $\|\tilde{A} - \tilde{B}\| < \epsilon$, where $B = B_1 + \cdots + B_k$.*

Proof The set

$$S = \{B \in M_3(\mathbb{R}) : B \neq 0, \|\tilde{A} - \tilde{B}\| < \varepsilon\}$$

is equal to the set

$$\{B \in M_3(\mathbb{R}) : B \neq 0, \angle(A, B) < \arccos(1 - \varepsilon^2/2)\}.$$

This is a cone. It is convex and closed under multiplication by a positive scalar. So, if $B_1, \ldots, B_k \in S$, then

$$B_1 + \cdots + B_k = k(k^{-1}B_1 + \cdots + k^{-1}B_k) \in S.$$

□

Example 2.66 Let us use the following notation: $X = (x, y, z)$. We first find a mapping $f : \overline{\mathbb{H}^3} \to \mathbb{R}^3$ such that

1. f is continuous on $\overline{\mathbb{H}^3}$.
2. f is not Lipschitz on $L = \{(0, 0, z) : 0 \leq z \leq 1\}$.
3. f is HQC on $\mathbb{H}^3$.

Note that the same is true for the restriction of f to the closed unit ball centered at $(0, 0, 1)$.

Let $g(X) = X/|X|^3$. This mapping is the gradient of a harmonic function $1/|X|$, up to a constant, and therefore harmonic for $X \neq 0$. Note that

$$Dg(X) = \frac{1}{|X|^3}(I - 3U_X^T \cdot U_X),$$

where $U_X = X/|X|$. Note also that $|g(X)| \leq 1/|X|^2$ and $\|Dg(X)\| \leq C_0/|X|^3$. Now set

$$f(X) = f_0(X) + \sum_{n=1}^{\infty} f_n(X),$$

where $f_0(X) = (x, y, -2z)$, $f_n(X) = c_n g(X - X_n)$, and $X_n = (0, 0, -r_n)$. We are going to show that the sequences r_n and c_n can be chosen so that the above three conditions are satisfied. We will require that they are strictly positive and that $\lim_{n\to\infty} r_n = 0$ monotonically.

First of all we impose the condition

$$\sum_{n=1}^{\infty} \frac{c_n}{r_n^2} < +\infty \qquad (C)$$

that will be sufficient for continuity up to the boundary. To see this note that, for every $X \in \overline{\mathbb{H}^3}$,

$$\sum_{n=1}^{\infty} |f_n(X)| = \sum_{n=1}^{\infty} \frac{c_n}{|X - X_n|^2} \leq \sum_{n=1}^{\infty} \frac{c_n}{r_n^2} < +\infty$$

and therefore the series $\sum_{n=1}^{\infty} f_n(X)$ converges absolutely and uniformly on $\overline{\mathbb{H}^3}$. Now we impose the condition

$$\sum_{n=1}^{\infty} \frac{c_n}{r_n^3} = +\infty \qquad (NL)$$

that is sufficient for the property 2. More precisely, for $X = (0, 0, z) \in \mathbb{H}^3$, we have

$$\frac{\partial}{\partial z} f_n(X) = -\frac{2c_n}{(z + r_n)^3} e_3,$$

where $e_3 = (0, 0, 1)$ and

$$\frac{\partial}{\partial z} f(X) = -\left[2 + \sum_{n=1}^{\infty} \frac{c_n}{(z + r_n)^3} \right] e_3.$$

It follows that if (NL) holds, then $|\frac{\partial}{\partial z} f(0, 0, z)| \to \infty$ as $z \to 0$ and this implies that f is not Lipschitz continuous on L.

Let us now look for a sufficient condition for quasiconformality. Towards this, note that a matrix $A \in M_3(\mathbb{R})$, $A \neq 0$ is K-quasiconformal iff its normalization

$\tilde{A} = A/\|A\|$ is K-quasiconformal. Moreover, for each compact $H \subset GL_3(\mathbb{R})$ there is a $K \geq 1$ such that every $A \in H$ is K-quasiconformal.

Let $A_0 = \mathrm{diag}(1, 1, -2) = Df_0$. Choose $0 < \delta < \sqrt{2}$ such that $H = \{B : \|\tilde{A}_0 - B\| \leq \delta\}$ is a compact subset of $GL_3(\mathbb{R})$. It follows that there is a $K_0 > 1$ such that every $B \in H$ is a K_0-quasiconformal matrix.

Let $A_X = I - 3U_X^T \cdot U_X$, $X \in \overline{\mathbb{H}^3}$, $X \neq 0$. Note that $A_X = A_{U_X}$. It is clear that A_X is continuous in X and $A_{(0,0,1)} = A_0$ and therefore there is an $\epsilon > 0$ such that $\|\tilde{A}_X - \tilde{A}_0\| < \delta$ whenever $|X - e_3| < \epsilon$. Putting it into an equivalent form, there is an $\eta > 0$ such that $\tan \angle(e_3, X) < \eta$ implies $\|\tilde{A}_X - \tilde{A}_0\| < \delta$ or, in coordinates, $\|\tilde{A}_X - \tilde{A}_0\| < \delta$ whenever $X = (x, y, z) \in \overline{\mathbb{H}^3}$ satisfies $\frac{\sqrt{x^2+y^2}}{z} < \eta$.

Now we choose $\beta > 0$ such that $\|A_0 - B\| \leq \beta$ implies $\|\tilde{A}_0 - \tilde{B}\| < \delta$.

Let us now show that the condition

$$2\sqrt{2}C_0 \sum_{r_n \leq \frac{\rho}{\eta}} \frac{c_n}{(r_n + \rho)^3} \leq \beta \quad \text{for all} \quad \rho > 0 \qquad (QC_1)$$

is sufficient for quasiconformality of f. Note that for $z > 0$, $Df(0, 0, z)$ is a constant multiple of A_0. Consider now $X = (x, y, z) \in \mathbb{H}^3$ with $\rho = \sqrt{x^2 + y^2} > 0$. Then

$$Df(X) = Df_0 + \sum_{n=1}^{\infty} Df_n(X)$$

$$= A_0 + \sum_{n=1}^{\infty} c_n A_{X-X_n}$$

$$= A_0 + \sum_{r_n \leq \frac{\rho}{\eta}} c_n A_{X-X_n} + \sum_{r_n > \frac{\rho}{\eta}} c_n A_{X-X_n}$$

$$= A_0 + R + T.$$

Note that the sum T is finite (possibly empty) and for each term in that sum we have

$$\frac{\sqrt{x^2 + y^2}}{z + r_n} \leq \frac{\rho}{r_n} \leq \eta.$$

It follows that $\|\tilde{A}_{X-X_n} - \tilde{A}_0\| < \delta$ for each term in that sum. Let us now estimate the norm of R:

$$\|R\| \leq \sum_{r_n \leq \frac{\rho}{\eta}} c_n \|A_{X-X_n}\|$$

$$\leq C_0 \sum_{r_n \leq \frac{\rho}{\eta}} \frac{c_n}{|X - X_n|^3}$$

$$= C_0 \sum_{r_n \le \frac{\ell}{\eta}} \frac{c_n}{\left[\rho^2 + (z + r_n)^2\right]^{3/2}}$$

$$\le C_0 \sum_{r_n \le \frac{\ell}{\eta}} \frac{c_n}{(\rho^2 + r_n^2)^{3/2}}$$

$$\le 2\sqrt{2} C_0 \sum_{r_n \le \frac{\ell}{\eta}} \frac{c_n}{(r_n + \rho)^3}$$

$$\le \beta.$$

It follows that $\|(A_0 + R) - A_0\| \le \beta$ and therefore $\|\tilde{A}_0 - \widetilde{A_0 + R}\| < \delta$. Note that $Df(X)$ can be represented as a sum of terms satisfying the assumptions of Lemma 2.65, and so $\|\widetilde{Df(X)} - \tilde{A}_0\| < \delta$, and $\widetilde{Df(X)} \in H$. It follows that $\widetilde{Df(X)}$ is a K_0-quasiconformal matrix and so is $Df(X)$. It is easy to verify that f is one-to-one, hence it follows that f is a K_0-quasiconformal map.

Note that the condition (QC_1) is equivalent to the following:

$$\sum_{k=n}^{\infty} c_k \le M r_n^3 \qquad \text{for all } n \ge 1 \qquad (QC),$$

where M is constant depending on η, C_0, and β. The exact value of M is not important for us here. This is so because once we have sequences r_n and c_n satisfying (C), (NL), and (QC) for some M, it suffices to multiply c_n with a suitably small constant to get M as small as desired. This follows from that fact that the conditions (C) and (NL) are invariant under such change of c_n.

Note that that the sequences $r_n = 2^{-2^n/3}$ and $c_n = 2^{-2^n}$ satisfy the conditions (C), (NL), and (QC). We can therefore conclude there is a non-Lipschitz qhc map on $\mathbb{B}^n$ continuous up to the boundary.

In conclusion we note that an analogous construction can be carried in all dimensions $k \ge 2$. Moreover, by multiplying the z-component of our function by a factor $-1/2$ and taking a tail of the series $\sum_{n=1}^{\infty} f_n(X)$, one can get the constant of quasiconformality as close to 1 as desired.

We would like to emphasize that the harmonicity is important here because it is easy to construct non-Lipschitz quasiconformal mappings.

2.7   Hölder Continuity of HQC Mappings

It is clear that for general quasiconformal mappings $u : \Omega_1 \to \Omega_2$ one cannot expect that the modulus of continuity behaves as in the Theorem 2.64, even for $\Omega_1 = \mathbb{B}^n$. However, for bounded Ω_2, Hölder continuity of $u|_{\partial\Omega_1}$ implies Hölder continuity of u, but with a possibly different Hölder exponent, see [120] and [111].

The following theorem is the main result of the paper [111].

Theorem 2.67 *Let D be a bounded domain in $\mathbb{R}^n$ and let f be a continuous mapping of $\overline{D}$ into $\mathbb{R}^n$ which is quasiconformal in D. Assume that for some $M > 0$ and $0 < \alpha \leq 1$,*

$$|f(x) - f(y)| \leq M|x - y|^{\alpha} \tag{2.26}$$

for x and y that lie on ∂D. Then

$$|f(x) - f(y)| \leq M'|x - y|^{\beta} \tag{2.27}$$

for all x and y on $\overline{D}$, where $\beta = \min(\alpha, K_I^{1/(1-n)})$ and M' depends only on M, α, n, $K(f)$, and $diam(D)$.

It turns out that the exponent β is the best possible, since there is an example of a radial quasiconformal map $f(x) = |x|^{\alpha-1}x$, $0 < \alpha < 1$, of $\overline{B}^n$ onto itself shows (see [155], p. 49). Moreover, the assumption of boundedness is essential. To see this, one can consider $g(x) = |x|^a x$, $|x| \geq 1$ where $a > 0$. Then g is quasiconformal in $D = \mathbb{R}^n \setminus \overline{B}^n$ (see [155], p. 49), it is identity on ∂D and hence Lipschitz continuous on ∂D. However, $|g(te_1) - g(e_1)| \asymp t^{a+1}$, $t \to \infty$, and therefore g is not globally Lipschitz continuous on D.

Regarding this result, P. Koskela has suggested the following question:

Question 2.68 Is it possible to replace β with α if we assume, in addition to quasiconformality, that f is harmonic?

This question was answered in [16] on which we base the rest of this section and where Theorem 2.76 ([16, Theorem 2.1]) below was proved showing that for a wide range of domains, including those with a uniformly perfect boundary, Hölder continuity on the boundary implies Hölder continuity with the same exponent inside the domain for the class of HQC mappings. In fact, we prove a more general result, including domains having a thin, (in the sense of capacity) portion of the boundary. However, this generality is in a sense illusory, because each HQC mapping extends harmonically and quasiconformally across such a portion of the boundary.

We need a notion of capacity. Recall that a condenser is a pair (K, U), where K is a nonempty compact subset of an open set $U \subset \mathbb{R}^n$. The capacity of the condenser (K, U) is defined as follows:

$$\mathrm{cap}(K, U) = \inf \int_{\mathbb{R}^n} |\nabla u|^n dV,$$

where infimum is taken over all continuous real-valued $u \in ACL^n(\mathbb{R}^n)$ such that $u(x) = 1$ for $x \in K$ and $u(x) = 0$ for $x \in \mathbb{R}^n \setminus U$. It turns out that the ACL^n condition can be replaced with the Lipschitz continuity in this definition. It should be noted that for a compact $K \subset \mathbb{R}^n$ and open bounded sets U_1 and U_2 containing K, one has: $\mathrm{cap}(K, U_1) = 0$ iff $\mathrm{cap}(K, U_2) = 0$. It follows that the notion of a compact set of zero capacity is well defined (see [158], Remarks 7.13), and that we can write $\mathrm{cap}(K) = 0$ in this realm. It turns out that the notions of modulus of a curve family defined earlier and capacity are related since, by results of [67] and [164], we have

$$\mathrm{cap}(K, U) = M(\Delta(K, \partial U; U)),$$

where $\Delta(E, F; G)$ denotes family of curves connecting E to F within G, see [155] or [158] for details.

Beside this notion of capacity related to quasiconformal mappings, we need also the notion of Wiener capacity related to harmonic functions. For a compact $K \subset \mathbb{R}^n$, $n \geq 3$, the Weiner capacity is defined by

$$\mathrm{cap}_W(K) = \inf \int_{\mathbb{R}^n} |\nabla u|^2 dV,$$

where infimum is taken over all Lipschitz continuous compactly supported functions u on $\mathbb{R}^n$ such that $u = 1$ on K. It should be noted that every compact $K \subset \mathbb{R}^n$ which has capacity zero has Wiener capacity zero. To see this, choose an open ball $B_R = B(0, R) \supset K$. Since $n \geq 2$ we have, by the Hölder inequality,

$$\int_{\mathbb{R}^n} |\nabla u|^2 dV \leq |B_R|^{1-2/n} \left(\int_{\mathbb{R}^n} |\nabla u|^n dV \right)^{2/n}$$

for each Lipschitz continuous u vanishing outside U, and our claim follows immediately from definitions.

The condenser capacity is a conformal invariant and behaves nicely under K-quasiconformal and K-quasiregular mappings, almost like a moduli of curve families (see [113, Theorem 6.2, Theorem 7.1]).

A compact set $K \subset \mathbb{R}^n$, consisting of at least two points, is α-uniformly perfect ($\alpha > 0$) if there is no ring R separating K (i.e., so that both components of $\mathbb{R}^n \setminus R$ intersect K) such that $\mathrm{mod}(R) > \alpha$. We say that a compact $K \subset \mathbb{R}^n$ is uniformly perfect if it is α-uniformly perfect for some $\alpha > 0$.

A characterization of uniform perfectness is given in the following theorem of P. Järvi and M. Vuorinen.

Theorem 2.69 ([71, Theorem 4.1]) *Let $E \subset \overline{\mathbb{R}^n}$ be a closed set containing at least two points. Then the following properties are equivalent:*

1. *E is α-uniformly perfect for some $\alpha > 0$;*
2. *there exist positive constants β and C_1 such that $\Lambda^\beta(\overline{B}^n(x, r) \cap E) \geqslant C_1 r^\beta$ for $x \in E \cap \mathbb{R}^n$ and $r \in (0, d(E))$;*
3. *there is a positive constant C_2 such that $\mathrm{cap}(x, E, r) \geqslant C_2$ for $x \in E \cap \mathbb{R}^n$ and $r \in (0, d(E))$.*

The constants α, β, C_1, and C_2 depend only on n and each other.

Uniformly perfect domains are important in Geometric Function Theory (see [80, Chapter 15], [47, pp. 343–345], [127]).

We denote the α-dimensional Hausdorff measure of a set $F \subset \mathbb{R}^n$ by $\Lambda_\alpha(F)$. If K is a compact set, then by [71, Corollary 4.2] $\Lambda^\beta(K)$ is positive for some positive β.

Let D denote a bounded domain in $\mathbb{R}^n$, $n \geq 3$. Let

$$\Gamma_0 = \{x \in \partial D : \operatorname{cap}(\overline{B}^n(x, \epsilon) \cap \partial D) = 0 \text{ for some } \epsilon > 0\},$$

and $\Gamma_1 = \partial D \setminus \Gamma_0$. Using this notation we can state the following facts which will be needed in the rest of this section.

Lemma 2.70 ([158, Lemma 7.14, p. 86] and [132, p. 72]) *Suppose that F is a compact set in $\mathbb{R}^n$ of capacity zero. Then for every $\alpha > 0$, the α-dimensional Hausdorff measure $\Lambda_\alpha(F)$ of F is zero. In particular, int $F = \emptyset$, and F is totally disconnected.*

Lemma 2.71 ([155, Theorem 35.1, p. 118]) *Suppose that $f : D \longrightarrow D'$ is a homeomorphism and that $E \subseteq D$ is a set such that E is closed in D and such that E has a σ-finite $(n-1)$-dimensional measure. Suppose also that every point in $D \setminus E$ has a neighborhood U such that $K_I(f|_U) \leq a$ and $K_O(f|_U) \leq b$. Then $K_I(f) \leq a$ and $K_O(f) \leq b$.*

Lemma 2.72 ([158, Corollary 5.41, p. 63]) *If $x \in \mathbb{R}^n$, $0 < a < b < \infty$, and $F_1, F_2 \subseteq B^n(x, a)$, $F_3 \subseteq \mathbb{R}^n \setminus B^n(x, b)$, $\Gamma_{ij} = \Delta(F_i, F_j)$, then*

1. $M(\Gamma_{12}) \geq 3^{-n} \min\{M(\Gamma_{13}), M(\Gamma_{23}), c_n(\log \frac{b}{a})\}$,
2. $M(\Gamma_{12}) \geq d(n, b/a) \min\{M(\Gamma_{13}) M(\Gamma_{23})\}$.

Lemma 2.73 ([111, Lemma 8]) *Let y be a point in a domain D, let x be a point in ∂D closest to y and let $d = |x - y|$. Suppose that*

$$\operatorname{cap}(\overline{B}^n(y, d/2), D) \geq m > 0.$$

If f is a continuous mapping of $\overline{D}$ into $\mathbb{R}^n$ which is quasiconformal in D and if the boundary mapping satisfies a Hölder condition at x with exponent α and constant M, then

$$|f(x) - f(y)| \leq \hat{M}|x - y|^\alpha,$$

where $\hat{M} > 0$ depends only on m, n, $K(f)$, α, and M.

Definition 2.74 We say that E is removable for the class H, where H is a class of functions harmonic in a bounded domain D, if it is possible to give metrical conditions on E which guarantee that every function in H can be extended to a harmonic function also in E.

Lemma 2.75 ([34, Theorem 2, p. 91]) *A set E is removable for the class H_α of harmonic functions satisfying a Lipschitz condition of order α, $0 < \alpha < 1$*

$$|u(x) - u(x')| \leq Const.|x - x'|^\alpha, \quad x, x' \in D,$$

if and only if $\Lambda_{d-2+\alpha}(E) = 0$.

Theorem 2.76 ([16]) *Let D be a bounded domain in $\mathbb{R}^n$ and assume $f : \overline{D} \to \mathbb{R}^n$ is continuous on $\overline{D}$ and harmonic and quasiconformal in D. Assume f is Hölder continuous with exponent α, $0 < \alpha \leq 1$, on ∂D and Γ_1 is uniformly perfect. Then f is Hölder continuous with exponent α on $\overline{D}$.*

If Γ_0 is empty, we obtain the following:

Corollary 2.77 ([16]) *Let D be a bounded domain in $\mathbb{R}^n$. If $f : \overline{D} \to \mathbb{R}^n$ is continuous on $\overline{D}$, Hölder continuous with exponent α, $0 < \alpha \leq 1$ on ∂D, harmonic and quasiconformal in D, and if ∂D is uniformly perfect, then f is Hölder continuous with exponent α on $\overline{D}$.*

The first step in the proof of Theorem 2.76 consists in reducing it to the case $\Gamma_0 = \emptyset$. In fact, we show that the existence of a HQC extension of f across Γ_0 follows from the well-known results. Let $D' = D \cup \Gamma_0$. Then D' is an open set in $\mathbb{R}^n$, Γ_0 is a closed subset of D', and $\partial D' = \Gamma_1$.

Note that $\mathrm{cap}(K \cap \Gamma_0) = 0$ for each compact $K \subset D'$, and therefore, by Lemma 2.70, $\Lambda_\alpha(K \cap \Gamma_0) = 0$ for each $\alpha > 0$. In particular, Γ_0 has σ-finite $(n-1)$-dimensional Hausdorff measure. Note that this set is closed in D', so we can apply Lemma 2.71 to conclude that f has a quasiconformal extension F across Γ_0, which has the same quasiconformality constant as f.

Our set Γ_0 is a countable union of compact sets K_j of capacity zero and, therefore, of Wiener capacity zero, so we conclude that Γ_0 has Wiener capacity zero. Therefore, by a Lemma 2.75 (see [34]), there is a (unique) extension $G : \overline{D'} \to \mathbb{R}^n$ of f which is harmonic in D'. Note that $F = G$ is a harmonic quasiconformal extension of f to $\overline{D'}$ which has the same quasiconformality constant as f.

Note that we have just reduced the proof of Theorem 2.76 to the proof of Corollary 2.77. So let us now start with the proof of Corollary 2.77. We need the following lemma.

Lemma 2.78 *Let $D \subset \mathbb{R}^n$ be a bounded domain with uniformly perfect boundary. There exists a constant $m > 0$ such that for every $y \in D$ we have*

$$cap(\overline{B}^n(y, \frac{d}{2}), D) \geq m , \qquad d = dist(y, \partial D). \tag{2.28}$$

Proof We start by fixing $y \in D$ as above and $z \in \partial D$ such that $|y - z| = d \equiv r$. Then $\mathrm{diam}(\partial D) = \mathrm{diam}(D) > 2r$. Set $F_1 = \overline{B}^n(z, r) \cap (\partial D)$ and $F_2 = \overline{B}^n(z, r) \cap \overline{B}^n(y, \frac{d}{2})$, $F_3 = S(z, 2r)$. Let $\Gamma_{i,j} = \Delta(F_i, F_j; \mathbb{R}^n)$ for $i, j = 1, 2, 3$. By Theorem 2.69 there exists a constant $a = a(E, n) > 0$ such that

$$M(\Gamma_{1,3}) \geq a,$$

while by standard estimates for the spherical ring (see Example 2.9) there exists $b = b(n) > 0$ such that

$$M(\Gamma_{2,3}) \geq b .$$

Now, by Lemma 2.72 there exists $m = m(E, n) > 0$ such that

$$M(\Gamma_{1,2}) \geq m .$$

As the final step, with $B = \overline{B}^n(y, d/2)$, we have

$$\mathrm{cap}(B, D) = M(\Delta(B, \partial D; \mathbb{R}^n)) \geq M(\Gamma_{1,2}) \geq m .$$

$\square$

Using this lemma, our assumption

$$|f(x_1) - f(x_2)| \leq C|x_1 - x_2|^\alpha , \qquad x_1, x_2 \in \partial D,$$

and Lemma 2.73, we conclude that there is a constant M, depending on $m, n, K(f)$, C, and α only such that

$$|f(x) - f(y)| \leq M|x - y|^\alpha , \quad y \in D, \ x \in \partial D, \ \mathrm{dist}(y, \partial D) = |x - y|.$$

It should be noted that an argument from [111] shows that the above estimate holds for $y \in D$, $x \in \partial D$ without any further conditions, but with possibly different constant:

$$|f(x) - f(y)| \leq M'|x - y|^\alpha , \quad y \in D, \ x \in \partial D. \tag{2.29}$$

The following lemma was proved in [32, Lemma A.1, p. 292] for real valued functions. The proof of this lemma relies on the maximum principle which holds also for vector valued harmonic functions, and so the lemma holds for harmonic mappings as well.

Lemma 2.79 *Assume $h : \overline{D} \to \mathbb{R}^n$ is continuous on $\overline{D}$ and harmonic in D, where D is bounded domain. Assume for each $x_0 \in \partial D$ we have*

$$\sup_{B^n(x_0,r) \cap D'} |h(x) - h(x_0)| \leq \omega(r) \qquad for \ 0 < r \leq r_0.$$

Then $|h(x) - h(y)| \leq \omega(|x - y|)$ whenever $x, y \in D$ and $|x - y| \leq r_0$.

Now we combine (2.29) and the above lemma, with $r_0 = \mathrm{diam}(D)$, to complete the proof of Corollary 2.77 and therefore of Theorem 2.76 as well.

Let ω is any nonnegative nondecreasing function satisfying $\omega(2t) \leq 2\omega(t)$ for $t \geq 0$. Let D be a bounded domain in $\mathbb{R}^n$, $n \geq 2$ and f a continuous mapping from $\overline{D}$ into $\mathbb{R}^n$ which is quasiconformal in D and satisfies that $|f(x) - f(y)| \leq \omega(|x-y|)$ for all $x, y \in \partial D$. In [17] it is shown that $|f(x) - f(y)| \leq C \max\{\omega(|x - y|), |x - y|^\alpha\}$ for all $x, y \in D$, where $\alpha = K_I(f)^{1/(1-n)}$.

2.8   Subharmonicity of the Modulus of HQR Mappings

In the paper [78] on which this section is based, Theorem 2.53 is extended to the n-dimensional setting and also Theorem 2.63 was improved by giving the optimal value of q. Moreover for the first time the case $q < 0$ was considered and the proofs in [78] are completely different from those presented above.

Let $0 < \lambda_1^2 \leqslant \lambda_2^2 \leqslant \cdots \leqslant \lambda_n^2$ be the eigenvalues of the matrix $Du(x)Du(x)^t$. Then

$$J_u(x) = \prod_{k=1}^{n} \lambda_k,$$

$$|Du| = \lambda_n$$

and

$$l(Du) = \lambda_1.$$

For the Hilbert–Schmidt norm of the matrix $Du(x)$ defined by

$$\|Du(x)\| = \sqrt{\mathrm{Trace}(Du(x)Du(x)^t)}$$

we have

$$\|Du(x)\| = \sqrt{\sum_{k=1}^{n} \frac{\partial u}{\partial x_k} \cdot \frac{\partial u}{\partial x_k}} = \sqrt{\sum_{k=1}^{n} \left| \frac{\partial u}{\partial x_k} \right|^2}$$

and

$$\|Du(x)\| = \sqrt{\sum_{k=1}^{n} \lambda_k^2}.$$

For a quasiregular mapping we have

$$\frac{\lambda_n}{\lambda_k}, \frac{\lambda_k}{\lambda_1} \leqslant K, \quad k = 1, \ldots, n.$$

Lemma 2.80 ([150, Chapter 7, 3.1.3, p. 218]) *Let $u : \Omega \longrightarrow \mathbb{R}^n$ be a harmonic mapping where Ω is a domain in $\mathbb{R}^n$. Then for $q \in \mathbb{R}$ and all $z \in \Omega \setminus u^{-1}(0)$*

$$\Delta |u(z)|^q = q \left[|u(z)|^{q-2} \|Du(z)\|^2 + (q-2)|u(z)|^{q-4} \left| \sum_{j=1}^{n} u_j(z)\nabla u_j(z) \right|^2 \right].$$
$$(2.30)$$

Proof Write $v := |u|^q = (u_1^2 + \cdots + u_n^2)^p$ for $p = q/2$. A direct computation gives

$$v_{x_i} = q(u_1^2 + \cdots + u_n^2)^{p-1}(u_1 u_{1 x_i} + \cdots + u_n u_{n x_i})$$

$$v_{x_i x_i} = 2q(p-1)(u_1^2 + \cdots + u_n^2)^{p-2}(u_1 u_{1 x_i} + \cdots + u_2 u_{2 x_i})^2$$
$$+ (u_1^2 + \cdots u_n^2)^{p-1}[u_1 u_{1 x_i x_i} + (u_1 u_{1 x_i})^2 + \cdots + u_n u_{1 x_i x_i} + (u_n u_{1 x_i})^2].$$

Therefore

$$\Delta v = q\{|u|^{q-2} \left[(u_1 \Delta u_1 + \cdots + u_n \Delta u_n) + \left(\sum_{k=1}^{n} u_{1 x_k}^2 + \cdots + \sum_{k=1}^{n} u_{1 x_k}^2 \right) \right]$$

$$+ (q-2)|u|^{q-4} \sum_{k=1}^{n} \sum_{j=1}^{n} u_j u_{j x_k}^2 \}$$

$$= q\{|u|^{q-2} \left(\sum_{k=1}^{n} u_{1 x_k}^2 + \cdots + u_{n x_k}^2 \right) + (q-2)|u|^{q-4} \sum_{k=1}^{n} \left(\sum_{j=1}^{n} u_j u_{j x_k} \right)^2 \}$$

$$= q|u|^{q-4}\{|u|^2 \sum_{j=1}^{n} \left(\sum_{k=1}^{n} u_j u_{j x_k}^2 \right) + (q-2) \sum_{k=1}^{n} \left(\sum_{j=1}^{n} u_j u_{j x_k} \right)^2 \}$$

$$= q|u|^{q-4}\{|u|^2 \sum_{j=1}^{n} |\nabla u_j|^2 + (q-2) \sum_{k=1}^{n} \left(\sum_{j=1}^{n} u_j u_{j x_k} \right)^2 \}.$$

$\square$

Lemma 2.81 ([78]) *For each $1 - (n-1)K^2 < q < 1 - \frac{n-1}{K^2}$ and $q \neq 0$ there is a (linear) harmonic K-quasiconformal mapping $u : \mathbb{R}^n \longrightarrow \mathbb{R}^n$ such that $|u|^q$ is not subharmonic.*

Proof Assume first that $q > 0$. We will consider the linear mapping $u : \mathbb{R}^n \longrightarrow \mathbb{R}^n$ defined by

$$u(x_1, \ldots, x_n) = (x_1, \ldots, x_{n-1}, x_n K) \qquad (2.31)$$

where $K \geqslant 1$. This is obviously harmonic and K-quasiconformal. Putting this into formula (2.30) we get

$$[(n-1) + K^2]|u|^2 + (q-2)\left|\sum_{j=1}^{n-1} x_j e_j + K^2 e_n x_n\right|^2 \geq 0$$

which is equivalent to

$$(n-1+K^2)\left[\sum_{n=1}^{j-1} x_j^2 + K^2 x_n^2\right] + (q-2)\left|\sum_{j=1}^{n-1} x_j e_j + K^2 e_n x_n\right|^2 \geq 0.$$

By choosing $x_1 = \cdots = x_{n-1} = 0$ and $x_n = 1$, we obtain

$$(n-1+K^2)K^2 \geq (2-q)K^4$$

which is equivalent to

$$q \geq 1 - \frac{n-1}{K^2}.$$

For $q < 0$ we consider the linear mapping $u : \mathbb{R}^n \longrightarrow \mathbb{R}^n$ defined by

$$u(x_1, \ldots, x_n) = (x_1, \ldots, x_{n-1}, x_n/K). \tag{2.32}$$

$\square$

Theorem 2.82 ([78]) *Let* $\Omega \subset \mathbb{R}^n$ *be a domain and let* $u : \Omega \to \mathbb{R}^n$ *be a K-quasiregular and harmonic mapping. Then the mapping* $g(x) = |u(x)|^q$ *is subharmonic in*

1. Ω *for* $q \geqslant \max\{1 - \frac{n-1}{K^2}, 0\}$;
2. $\Omega \setminus u^{-1}(0)$, *for* $q \leqslant 1 - (n-1)K^2$.

Moreover, for $1 - (n-1)K^2 < q < 1 - \frac{n-1}{K^2}$ *and* $q \neq 0$ *there exists a K-quasiconformal harmonic mapping such that* $|u|^q$ *is not subharmonic.*

Proof Let us fix such a map $u : \Omega \to \mathbb{R}^n$ and set $\Omega_0 = \Omega \setminus u^{-1}\{0\}$. We have to find all positive real numbers q such that $\Delta|u|^q \geq 0$ on Ω_0. Since u is quasiregular, the set $Z = \{x \in \Omega_0 : \det Du(x) = 0\}$ has measure zero (see [158]). The set Z is also closed since u is smooth. In particular, $\Omega_1 = \Omega_0 \setminus Z$ is dense in Ω_0 and thus it suffices to prove that $\Delta|u|^q \geq 0$ on Ω_1.

By Lemma 2.80, we find all real q such that

$$q \left(|u|^{q-2} \|Du\|^2 + (q-2)|u|^{q-4} \left| \sum_{j=1}^{n} u_j \nabla u_j \right|^2 \right) \geq 0.$$

If $q \geq 2$, then $\Delta |u|^q \geq 0$. Assume that $q \geq 0$ and $q < 2$ such that

$$\left| \sum_{j=1}^{n} u_j(x) \nabla u_j(x) \right|^2 \leq \frac{1}{2-q} |u(x)|^2 \|Du(x)\|^2, \quad x \in \Omega_1.$$

After normalization, we see that it suffices to find a constant $q = q(K,n) < 2$ such that

$$\sup \left| \sum_{j=1}^{n} z_j \nabla u_j(x) \right|^2 \leq \frac{1}{2-q} \|Du(x)\|^2, \quad x \in \Omega_1. \tag{2.33}$$

Let $0 < \lambda_1^2 \leq \lambda_2^2 \leq \cdots \leq \lambda_n^2$ be the eigenvalues of the matrix $Du(x)Du(x)^t$. Then

$$\sup_{z \in S^{n-1}} \left| \sum_{j=1}^{n} z_j \nabla u_j(x) \right|^2 = \lambda_n^2 \tag{2.34}$$

$$\inf_{z \in S^{n-1}} \left| \sum_{j=1}^{n} z_j \nabla u_j(x) \right|^2 = \lambda_1^2 \tag{2.35}$$

and

$$\|Du(x)\|^2 = \sum_{k=1}^{n} \lambda_k^2. \tag{2.36}$$

Because u is K-quasiregular, we have

$$\frac{\lambda_n}{\lambda_k} \leq K, \quad k = 1, \ldots, n-1. \tag{2.37}$$

Thus (2.33) can be written as

$$\lambda_n^2 \leq \frac{1}{2-q} \sum_{k=1}^{n} \lambda_k^2. \tag{2.38}$$

By (2.36) and (2.37) we get that the inequality (2.38) is satisfied whenever

$$\frac{1}{1 + \frac{n-1}{K^2}} \leq \frac{1}{2 - q} \tag{2.39}$$

i.e.,

$$\max\{0, 1 - \frac{n-1}{K^2}\} \leq q < 2. \tag{2.40}$$

If $q < 0$, then we should have

$$\inf_{z \in S^{n-1}} \left| \sum_{j=1}^{n} z_j \nabla u_j(x) \right|^2 \geq \frac{1}{2 - q} \|Du(x)\|^2, \quad x \in \Omega_1, \tag{2.41}$$

i.e.,

$$2 - q \geq \sum_{k=1}^{n} \frac{\lambda_k^2}{\lambda_1^2}.$$

Because u is K-quasiregular

$$\frac{\lambda_k}{\lambda_1} \leq K, \quad k = 2, \ldots, n. \tag{2.42}$$

Thus if

$$q \leq 1 - (n - 1)K^2, \tag{2.43}$$

then (2.41) holds.

By Lemma 2.81, we only need to take $\tilde{u} = u|_\Omega$, where u is defined in (2.31) respectively in (2.32).   □

Remark 2.83 If $n = 2$, $1 - \frac{n-1}{K^2} = 1 - K^{-2}$. Thus, Theorem 2.82 is n-extension of Theorem 2.53.

Remark 2.84 In the case $1 \leqslant K \leqslant \sqrt{n-1}$, the function $|u|^q$ is subharmonic for all $q > 0$.

Chapter 3
Hyperbolic Type Metrics

The natural setup for our work here is a metric space (G, m_G) where G is a subdomain of $\mathbb{R}^n$, $n \geq 2$. For our studies, the distance $m_G(x, y), x, y \in G$ is required to take into account both how close the points x, y are to each other and the position of the points relative to the boundary ∂G. Metrics of this type are called hyperbolic type metrics and they are substitutes for the hyperbolic metric in dimensions $n \geq 3$. The quasihyperbolic metric and the distance ratio metric are both examples of hyperbolic type metrics. A key problem is to study a quasiconformal mapping between metric spaces

$$f : (G, m_G) \to (f(G), m_{f(G)})$$

and to estimate its modulus of continuity. We expect Holder continuity, but a concrete form of these results may differ from metric to metric. Another question is the comparison of the metrics to each other.

In this chapter we will introduce the following types of metrics:

1. Chordal metric q on $\overline{\mathbb{R}^n} = \mathbb{R}^n \cup \{\infty\}$.
2. Hyperbolic metric ρ_D on $\mathbb{B}^n$ or $\mathbb{H}^n$.
3. Quasihyperbolic metric k_D of a domain $D \subset \mathbb{R}^n$.
4. A metric j_D closely related to k_D.

The last two are defined in all proper subdomains $D \subset \mathbb{R}^n$. Both of them generalize hyperbolic metric (on $\mathbb{B}^n$ or $\mathbb{H}^n$) to arbitrary proper subdomain $D \subset \mathbb{R}^n$. If $D = \mathbb{H}^n$, then $k_D = \rho$, but the quasihyperbolic and hyperbolic metrics are not equal on the unit ball. This chapter is based on [86] and [98].

© Springer Nature Switzerland AG 2019   57
V. Todorčević, *Harmonic Quasiconformal Mappings and Hyperbolic Type Metrics*,
https://doi.org/10.1007/978-3-030-22591-9_3

3.1 Möbius Transformations

Before discussing these metrics, we make a brief excursion into the theory of Möbius transformations [24].

We define the inversion with respect to the sphere $S^{n-1}(a, r)$ by the formula

$$\varphi(x) = a + \left(\frac{r}{|x - a|} \right)^2 (x - a). \qquad (3.1)$$

In the special case of the unit sphere, S^{n-1}, this becomes $\varphi(x) = \frac{x}{|x|^2}$. We also write $x \mapsto x^*$, where $x^* = \frac{x}{|x|^2}$. Then (3.1) becomes

$$\varphi(x) = a + r^2 (x - a)^*. \qquad (3.2)$$

We extend φ by continuity to $\overline{\mathbb{R}^n}$: $\varphi(a) = \infty$, $\varphi(\infty) = a$.

Note that $\varphi^2(x) = (\varphi \circ \varphi)(x) = x$ for all $x \in \overline{\mathbb{R}^n}$ and $\varphi(x) = x \Leftrightarrow x \in S^{n-1}(a, r)$.

Lemma 3.1 *If $a \in \mathbb{R}^n$, $r > 0$ and φ is inversion with respect to the sphere $S^{n-1}(a, r)$, then for all $x, y \in \mathbb{R}^n$, we have*

$$|\varphi(x) - \varphi(y)| = \frac{r^2 |x - y|}{|x - a| \, |y - a|}. \qquad (3.3)$$

Proof

$$|\varphi(x) - \varphi(y)| = |(\varphi(x) - a) - (\varphi(y) - a)| = |r^2(x-a)^* - r^2(y-a)^*|$$

$$= r^2 \left| \frac{x - a}{|x - a|^2} - \frac{y - a}{|y - a|^2} \right|$$

$$= r^2 \sqrt{ \frac{|x - a|^2}{|x - a|^4} + \frac{|y - a|^2}{|y - a|^4} - 2 \frac{(x-a)(y-a)}{|x-a|^2 |y-a|^2} }$$

$$= r^2 \sqrt{ \frac{1}{|x - a|^2} + \frac{1}{|y - a|^2} - 2 \frac{(x-a)(y-a)}{|x-a|^2 |y-a|^2} }$$

$$= \frac{r^2}{|x - a| \, |y - a|} \sqrt{ |y - a|^2 + |x - a|^2 - 2(x-a)(y-a) }$$

$$= \frac{r^2 |x - y|}{|x - a| \, |y - a|}.$$

$\square$

We also consider hyperplanes $P(a, t) = \{ x \in \mathbb{R}^n \mid \langle x, a \rangle = t \} \cup \{\infty\}$, where $a \in \mathbb{R}^n$, $a \neq 0$, and t is a real number.

A reflection in $P(a, t)$ is a map φ defined as follows:

$$\varphi(x) = x - 2[\langle x, a \rangle - t] \cdot a^*, \qquad x \in \mathbb{R}^n. \qquad (3.4)$$

We have that $\varphi^2(x) = x$ for $x \in \overline{\mathbb{R}^n}$ and $\varphi(x) = x$ for $x \in P(a, t)$.

Definition 3.2 A *Möbius transformation* in $\overline{\mathbb{R}^n}$ is a finite composition of inversions (with respect to spheres or hyperplanes).

Möbius transformations form a group, because $\varphi^2 = id$ for each inversion φ. We denote the group of Möbius transformations by $\mathscr{GM}(\overline{\mathbb{R}^n})$.

Definition 3.3 ([158, p. 3]) A map f in $\mathscr{GM}(\overline{\mathbb{R}^n})$ with $f(\infty) = \infty$ is called a *similarity* transformation if $|f(x) - f(y)| = c|x - y|$ for all $x, y \in \mathbb{R}^n$ where c is a positive number.

Theorem 3.4 ([4, p. 19]) *Each Möbius transformation is conformal.*

Proof It suffices, by the chain rule, to prove that each inversion is conformal. This is geometrically obvious for reflections with respect to hyperplanes. It remains to consider reflections in spheres. We shall consider the case of the unit sphere. The general case can be treated similarly.

Set $f(x) = x^*$. Then

$$\frac{\partial f_i}{\partial x_j}(x) = \frac{1}{|x|^2}\left(\delta_{ij} - \frac{2x_i x_j}{|x|^2}\right), \qquad (3.5)$$

and therefore

$$f'(x) = \frac{1}{|x|^2}(I - 2Q(x)),$$

where

$$Q(x) = \left(\frac{x_i x_j}{|x|^2}\right)_{i,j=1}^{n}$$

is a symmetric matrix (i.e., $Q = Q^T$) and $Q^2 = Q$. Then, we have the following:

$$f'(x) \cdot f'(x)^T = \frac{1}{|x|^4}(I - 2Q(x))^2 = \frac{1}{|x|^4}(I - 4Q(x) + 4Q^2(x)) = \frac{1}{|x|^4}I.$$

So, $|x|^2 f'(x)$ is an orthogonal matrix, hence $f'(x)$ is a conformal matrix. □

Note that since each inversion f changes orientation, it can be shown (see [129, pp. 137–145]) that we have the inequality

$$det(f'(x)) < 0.$$

3.2 Chordal Metric

The chordal metric q is defined by (Fig. 3.1)

$$q(x, y) = |\pi(x) - \pi(y)|; \quad x, y \in \overline{\mathbb{R}}^n, \tag{3.6}$$

where the stereographic projection $\pi : \overline{\mathbb{R}}^n \longrightarrow S^n(e_{n+1}/2, 1/2)$ is defined by

$$\pi(x) = e_{n+1} + \frac{x - e_{n+1}}{|x - e_{n+1}|^2}, \quad x \in \mathbb{R}^n, \quad \pi(\infty) = e_{n+1}. \tag{3.7}$$

From (3.7) and Lemma 3.1 follows

$$q(x, y) = \begin{cases} \dfrac{|x - y|}{\sqrt{1 + |x|^2}\sqrt{1 + |y|^2}}, & x \neq \infty \neq y, \\ \dfrac{1}{\sqrt{1 + |x|^2}}, & y = \infty. \end{cases} \tag{3.8}$$

Let (X_1, d_1) and (X_2, d_2) be metric spaces and let $f : X_1 \to X_2$ be a homeomorphism. We call f an isometry if $d_2(f(x), f(y)) = d_1(x, y)$ for all $x, y \in X_1$. A map f in $\mathscr{GM}(\overline{\mathbb{R}}^n)$ is called a spherical isometry if $q(f(x), f(y)) = q(x, y)$ for all $x, y \in \overline{\mathbb{R}}^n$.

An orthogonal transformation preserves the chordal metric because $|x|, |y|$ and $|x - y|$ are invariant under orthogonal transformations. The same is true for inversion with respect to the unit sphere S^{n-1}.

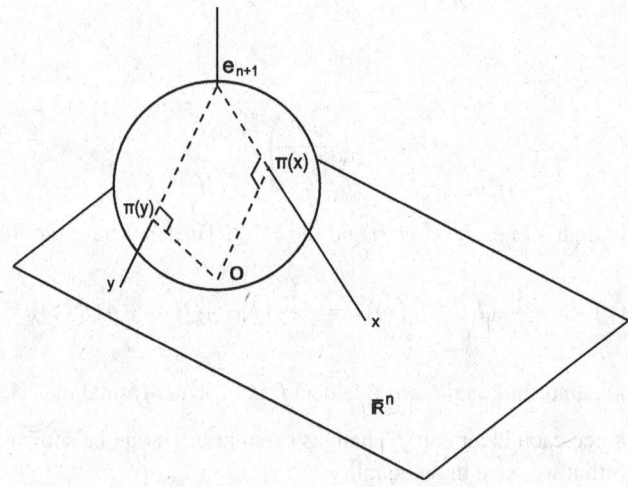

Fig. 3.1 Formulas (3.6) and (3.7) visualized [158, Diagram 1.1, p. 5]

Lemma 3.5 (Symmetry Lemma) *For all nonzero x and y in $\mathbb{R}^n$,*

$$\left| \frac{y}{|y|} - |y|x \right| = \left| \frac{x}{|x|} - |x|y \right|.$$

Lemma 3.6 *The inversion with respect to the unit sphere S^{n-1} preserves the chordal metric.*

Proof Using the symmetry lemma, we have

$$q(f(x), f(y)) = q(x^*, y^*) = \frac{|x^* - y^*|}{\sqrt{1 + |x^*|^2}\sqrt{1 + |y^*|^2}} = \frac{\left| \frac{x}{|x|^2} - \frac{y}{|y|^2} \right|}{\sqrt{1 + \frac{1}{|x|^2}}\sqrt{1 + \frac{1}{|y|^2}}}$$

$$= \frac{||y|^2 x - |x|^2 y|}{|x|^2|y|^2} \cdot \frac{1}{\sqrt{\frac{(1 + |x|^2)(1 + |y|^2)}{|x|^2|y|^2}}} = \frac{\frac{||y|^2 x - |x|^2 y|}{|x||y|}}{\sqrt{1 + |x|^2}\sqrt{1 + |y|^2}}$$

$$= \frac{\left| |y|\frac{x}{|x|} - |x|\frac{y}{|y|} \right|}{\sqrt{1 + |x|^2}\sqrt{1 + |y|^2}} = \frac{|x - y|}{\sqrt{1 + |x|^2}\sqrt{1 + |y|^2}} = q(x, y).$$

$\square$

We now present some simple properties of the chordal metric q.

Lemma 3.7 ([158, Exercise 1.18 (1), p. 6]) *For $0 < t < 1$ let $\omega(t) = t/\sqrt{1 - t^2}$, we have that $q(0, \omega(t) e_1) = t$ and that*

$$\frac{t}{s} \leqslant \frac{\omega(t)}{\omega(s)} \leqslant \frac{2t}{s}$$

for $0 < s < t < \frac{\sqrt{3}}{2}$.

Proof Let us only show the above equality since the proof of both inequalities is simple. Let $A = e_{n+1}$, $B = \omega(t) e_1$, $C = 0$, and $D = \pi(B)$. Since lines AB and CD are orthogonal, the triangles $\triangle ABC$ and $\triangle ACD$ are similar. Then,

$$q(0, \omega(t) e_1) = CD = \frac{CD}{1} = \frac{CD}{AC} = \frac{BC}{AB} = \frac{\frac{t}{\sqrt{1-t^2}}}{\sqrt{1 + \left(\frac{t}{\sqrt{1-t^2}}\right)^2}} = t.$$

$\square$

Lemma 3.8 ([158, Exercise 1.18 (2), p. 6]) *Let* $x, y \in \mathbb{B}^n$ *with* $s = q(0, x)$ *and* $t = q(0, y)$. *Then*

$$q(x, y) \leqslant s\sqrt{1 - t^2} + t\sqrt{1 - s^2} \leqslant t + s.$$

Proof

$$q(x, y) = \frac{|x - y|}{\sqrt{1 + |x|^2}\sqrt{1 + |y|^2}} \leqslant \frac{|x| + |y|}{\sqrt{1 + |x|^2}\sqrt{1 + |y|^2}}$$

$$= \frac{s}{\sqrt{1 + |y|^2}} + \frac{t}{\sqrt{1 + |x|^2}} \leqslant s + t.$$

The inequality follows from

$$\sqrt{1 - t^2} = \sqrt{1 - \frac{|y|^2}{1 + |y|^2}} = \frac{1}{\sqrt{1 + |y|^2}}.$$

$\square$

Lemma 3.9 ([158, Exercise 1.18 (4), p. 6]) *For* $x, y, z \in \mathbb{B}^n$, $x \neq z$, *we have*

$$\frac{1}{\sqrt{2}} \frac{|x - y|}{|x - z|} \leqslant \frac{q(x, y)}{q(x, z)} \leqslant \sqrt{2} \frac{|x - y|}{|x - z|}.$$

Proof This follows from

$$\frac{q(x, y)}{q(x, z)} = \frac{|x - y|}{\sqrt{1 + |x|^2}\sqrt{1 + |y|^2}} : \frac{|x - z|}{\sqrt{1 + |x|^2}\sqrt{1 + |z|^2}} = \frac{|x - y|}{|x - z|}\sqrt{\frac{1 + |z|^2}{1 + |x|^2}}$$

and

$$\frac{1}{\sqrt{2}} \leqslant \sqrt{\frac{1 + |z|^2}{1 + |x|^2}} \leqslant \sqrt{2},$$

because $x, z \in \mathbb{B}^n$.

$\square$

The balls $Q(x, r)$ in q-metrics are in fact Euclidean balls, half-spaces, or complements of Euclidean balls, corresponding respectively to the cases $\infty \notin Q(x, r)$, $\infty \in \partial Q(x, r)$, and $\infty \in Q(x, r)$.

$$x \in Q(x_0, r) \Leftrightarrow \frac{|x - x_0|}{\sqrt{1 + |x_0|^2}\sqrt{1 + |x|^2}} \leqslant r \Leftrightarrow |x - x_0|^2 \leqslant r^2(1 + |x_0|^2)(1 + |x|^2).$$

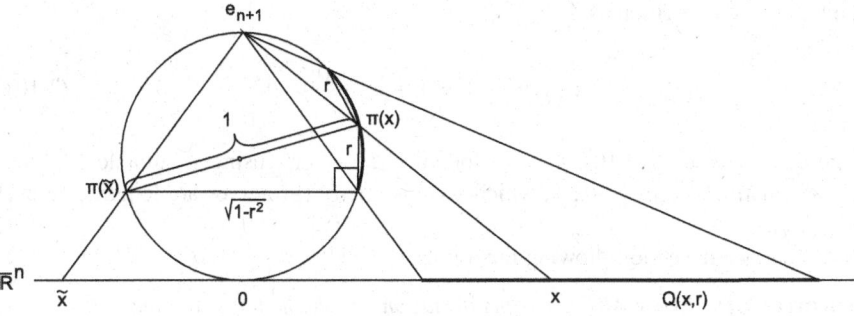

Fig. 3.2 A cross-section of the Riemann sphere [158, Diagram 1.3, p. 7]

Let $c = r^2(1 + |x_0|^2)$. Then

$$|x|^2 - 2\langle x, x_0 \rangle + |x_0|^2 \leqslant c|x|^2 + c \Leftrightarrow (c-1)|x|^2 - 2\langle x, x_0 \rangle + c - |x_0|^2 \leqslant 0.$$

The last inequality defines one of the sets mentioned above (Fig. 3.2).

3.3 Hyperbolic Metric

There are two domains in which hyperbolic geometry is developed in case $n \geq 3$. These are the unit ball $\mathbb{B}^n$ and the Poincare half-space $\mathbb{H}^n = \{x \in \mathbb{R}^n : x_n > 0\}$.

We start with $\mathbb{H}^n$ and define arc length element $ds = \frac{|dx|}{x_n}$. So the hyperbolic length of a curve $\gamma : [a, b] \to \mathbb{H}^n$ is defined by

$$l_h(\gamma) = \int_\gamma ds = \int_a^b \frac{|\gamma'(t)|\, dt}{\gamma_n(t)}.$$

Hyperbolic distance $\rho(a, b)$ between two points $a, b \in \mathbb{H}^n$ is defined as

$$\inf l_h(\gamma),$$

where inf is taken over all rectifiable curves γ joining a and b. In fact this infimum is attained by a length minimizing curve called a geodesic.

In the special case when one point is above the other (on the same vertical line; see [24, p.135]), the only geodesic arc connecting these points is line segment $[a, b] = \{(1 - t)a + tb : 0 \le t \le 1\}$. If $a = r\, e_n$, $b = s\, e_n$ and $r, s > 0$, we see that

$$\rho(a, b) = \rho(re_n, se_n) = \left| \int_r^s \frac{dt}{t} \right| = \left| \ln \frac{r}{s} \right| \quad \text{(by a simple integration).} \qquad (3.9)$$

This can also be written as

$$\cosh \rho(a, b) = 1 + \frac{|a - b|^2}{2rs}. \tag{3.10}$$

One can show that (3.10) is valid for all $a, b \in \mathbb{H}^n$, using a suitable Möbius transformation $\varphi : \mathbb{H}^n \to \mathbb{H}^n$, which is an isometry (preserves arc length element $|dx|/x_n$).

In fact, we have the following theorem:

Theorem 3.10 *Every Möbius transformation φ which maps $\mathbb{H}^n$ onto $\mathbb{H}^n$ is an isometry with respect to the hyperbolic metric.*

We give an outline of the proof of this important fact.

In view of formula (3.10) it suffices to show that the expression

$$\frac{|x - y|^2}{x_n \, y_n}, \quad x, y \in \mathbb{H}^n$$

is invariant under the group $\mathscr{GM}(\mathbb{H}^n)$.

For this, one has to consider inversions with respect to spheres $S^{n-1}(a, r)$, $a \in \partial \mathbb{H}^n$. The case of reflections in hyperplanes orthogonal to $\partial \mathbb{H}^n$ is trivial. (Note that inversion in $S^{n-1}(a, r)$ preserves $\mathbb{H}^n$ iff $a_n = 0$.)

So, let φ be such inversion. Then, for $x, y \in \mathbb{H}^n$ we have

$$\frac{|\varphi(x) - \varphi(y)|^2}{\varphi(x)_n \, \varphi(y)_n} = \left(\frac{r^2 |x - y|}{|x - a| \, |y - a|} \right)^2 \cdot \frac{|x - a|^2}{r^2 \, x_n} \cdot \frac{|y - a|^2}{r^2 \, y_n} = \frac{|x - y|^2}{x_n \, y_n},$$

where the first equality follows from (3.3) and the second equality from the fact that

$$\varphi(x)_n = 0 + \frac{r^2 \, x_n}{|x - a|^2}.$$

Since φ preserves angles and the ray $\{r \, e_n \, | \, r > 0\}$ is orthogonal to $\partial \mathbb{H}^n$, it follows that geodesics are semicircles orthogonal to $\partial \mathbb{H}^n$ (Fig. 3.3).

Absolute (cross) ratio of an ordered quadruple a, b, c, d of distinct points in $\overline{\mathbb{R}^n}$ is defined by

$$|a, b, c, d| = \frac{q(a, c) \, q(b, d)}{q(a, b) \, q(c, d)}. \tag{3.11}$$

Using our formula for $q(x, y)$ we see that for $a, b, c, d \in \mathbb{R}^n$, we have

$$|a, b, c, d| = \frac{|a - c| \cdot |b - d|}{|a - b| \cdot |c - d|}. \tag{3.12}$$

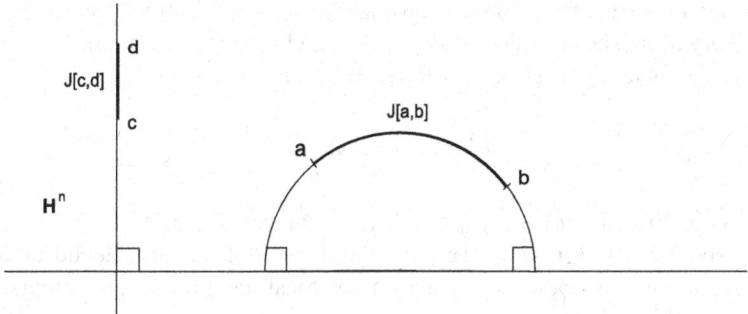

Fig. 3.3 Some geodesics of $\mathbb{H}^n$ [158, Diagram 2.1, p. 20]

This notion is important in the study of Möbius transformation, and the main reason for this is the following theorem.

Theorem 3.11 ([24, Theorem 3.2.7]) *A mapping* $\varphi : \overline{\mathbb{R}^n} \to \overline{\mathbb{R}^n}$ *is a Möbius transformation iff it preserves absolute ratios.*

We need only the "only if" part of this theorem. It is clear from (3.12) that Euclidean isometries, or more generally similarity mappings, preserve absolute ratio.

So, we have to show that inversion $\varphi(x) = x^*$ preserves absolute ratios.

However,

$$|x^* - y^*| = \frac{|x - y|}{|x| \cdot |y|}$$

and invariance again follows from (3.12).

Note that the order of a, b, c, d is important. In fact,

$$|0, e_1, x, \infty| = |x| = \frac{1}{|0, x, e_1, \infty|},$$

$$|0, e_1, \infty, x| = |x - e_1| = \frac{1}{|0, \infty, e_1, x|},$$

$$|0, \infty, x, e_1| = \frac{|x|}{|x - e_1|} = \frac{1}{|0, x, \infty, e_1|}.$$

Let $z, w \in \mathbb{H}^n$ and let L be an arc of a circle perpendicular to $\partial \mathbb{H}^n$ with $z, w \in L$ and let $\{z_*, w_*\} = L \cap \partial \mathbb{H}^n$, the points being labeled so that z_*, z, w, w_* occur in this order on L.

Now we prove that

$$\rho(z, w) = \log |z_*, z, w, w_*|. \tag{3.13}$$

We already know that both sides are invariant under $\varphi \in \mathscr{GM}(\mathbb{H}^n)$, so we can use an auxiliary Möbius map which sends z to $s\,e_n$ and w to $t\,e_n$ ($s, t > 0$).

We can assume $s < t$. Then $z_* = 0$, $w_* = \infty$, and

$$|0, s\,e_n, t\,e_n, \infty| = \frac{t}{s},$$

and we know that $\rho(z, w) = \log \frac{t}{s}$ and that is sufficient (Fig. 3.4).

Hyperbolic balls $D(a, M) = \{x \in \mathbb{H}^n \mid \rho(a, x) < M\}$ are also Euclidean balls. The balls of the same radii with center on a vertical line fill in a cone around that vertical line (Fig. 3.5).

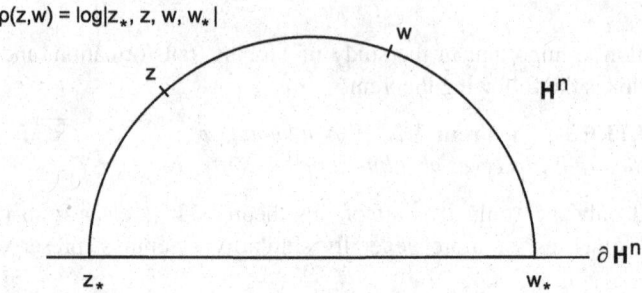

Fig. 3.4 The quadruple z_*, z, w, w_* [158, Diagram 2.3, p. 22]

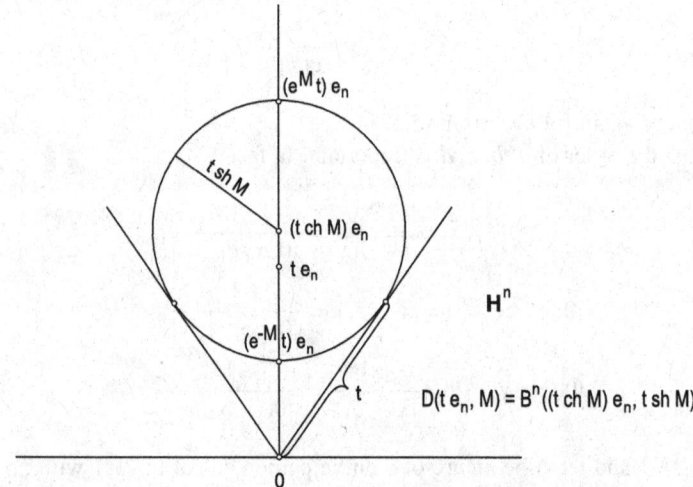

Fig. 3.5 The hyperbolic ball $D(t\,e_n, M)$ as an Euclidean ball [158, Diagram 2.4, p. 22]

3.4 Distance Ratio Metric j_D

For an open set $D \subset \mathbb{R}^n$, $D \neq \mathbb{R}^n$ we define $d(z) = d(z, \partial D)$ for $z \in D$ and

$$j_D(x, y) = \log\left(1 + \frac{|x - y|}{\min\{d(x), d(y)\}}\right) \qquad (3.14)$$

for $x, y \in D$. It is immediate that j_D has an invariant property in the sense that $j_D(x, y) = j_{f(D)}(f(x), f(y))$ if mapping f is a similarity.

For a nonempty $A \subset D$, we define the j_D-diameter of A by

$$j_D(A) = \sup\{j_D(x, y) \mid x, y \in A\}.$$

Now, we show that j_D is a metric on D.

Obviously, (M1), (M2), and (M3) hold for j_D. The proof of (M4) is divided into three cases:

1. In the case $\min\{d(x), d(y), d(z)\} = d(y)$, (M4) follows from

$$\left(1 + \frac{|x - y|}{d(y)}\right)\left(1 + \frac{|y - z|}{d(y)}\right) \geqslant 1 + \frac{|x - y|}{d(y)} + \frac{|y - z|}{d(y)} \geqslant 1 + \frac{|x - z|}{d(y)}.$$

2. In the case that $d(y)$ is between $d(x)$ and $d(z)$, we can assume that $d(x) \leqslant d(z)$ because the other case is analogous. We have to prove that

$$\left(1 + \frac{|x - y|}{d(x)}\right)\left(1 + \frac{|y - z|}{d(y)}\right) \geqslant 1 + \frac{|x - z|}{d(x)}.$$

This is equivalent to

$$\frac{|y - z|}{d(y)} + \frac{|x - y| \cdot |y - z|}{d(x)\,d(y)} \geqslant \frac{|x - z| - |x - y|}{d(x)}.$$

Because $|z - y| \geqslant |x - z| - |x - y|$, it is enough to prove that

$$\frac{|y - z|}{d(y)} + \frac{|x - y| \cdot |y - z|}{d(x)\,d(y)} \geqslant \frac{|z - y|}{d(x)}.$$

The last formula is equivalent to

$$\frac{|y - z|}{d(x)\,d(y)}(|x - y| - (d(y) - d(x))) \geqslant 0,$$

which is true, because $|x - y| \leqslant d(y) - d(x)$ holds for all $x, y \in \mathbb{R}^n$.

3. If $d(y) = \max\{d(x), d(y), d(z)\}$, we may assume that $d(x) \leqslant d(z)$, because the other case is analogous. We have to prove that

$$\left(1 + \frac{|x-y|}{d(x)}\right)\left(1 + \frac{|y-z|}{d(z)}\right) \geqslant 1 + \frac{|x-z|}{d(x)}.$$

This is equivalent to

$$\frac{|y-z|}{d(z)} + \frac{|x-y| \cdot |y-z|}{d(x)\,d(z)} \geqslant \frac{|x-z| - |x-y|}{d(x)}.$$

Since $|z - y| \geqslant |x - z| - |x - y|$, it is enough to prove that

$$\frac{|y-z|}{d(z)} + \frac{|x-y| \cdot |y-z|}{d(x)\,d(z)} \geqslant \frac{|z-y|}{d(x)}.$$

The last formula follows from

$$\frac{|y-z|}{d(x)\,d(z)}(|x-y| - (d(z) - d(x))) \geqslant \frac{|y-z|}{d(x)\,d(z)}(|x-y| - (d(y) - d(x))) \geqslant 0.$$

In the following lemma, we note some simple properties of j_D.

Lemma 3.12 ([158, Lemma 2.36, p. 28]) *For all* $x, y \in D$ *the following inequalities hold:*

1. $j_D(x, y) \geqslant |\log \frac{d(x)}{d(y)}|.$
2. $j_D(x, y) \leqslant |\log \frac{d(x)}{d(y)}| + \log\left(1 + \frac{|x-y|}{d(x)}\right) \leqslant 2\,j_D(x, y).$

Proof

1. Taking infimum over $z \in \partial D$ in the inequality $|y - z| \leqslant |x - z| + |x - y|$, we conclude that $d(y) \leqslant d(x) + |x - y|$.
2. If $d(x) \leqslant d(y)$, then the first inequality follows from the definition of j_D, and the second one from 1) and definition.

    Now, let $d(x) \geqslant d(y)$. Then,

$$j_D(x, y) = \log\left(1 + \tfrac{|x-y|}{d(y)}\right) \leqslant \log\left(\tfrac{d(x)}{d(y)} + \tfrac{d(x)}{d(y)} \cdot \tfrac{|x-y|}{d(y)}\right)$$
$$= \left|\log \tfrac{d(x)}{d(y)}\right| + \log\left(1 + \tfrac{|x-y|}{d(y)}\right) \leqslant 2\,j_D(x, y). \qquad (3.15)$$

The last inequality follows from (1).

□

For an open set $D \subset \mathbb{R}^n$, $D \neq \mathbb{R}^n$, and a nonempty $A \subset D$ such that $d(A, \partial D) > 0$ we define

$$r_D(A) = \frac{d(A)}{d(A, \partial D)}.$$

Now we prove the following inequalities:

$$\frac{1}{2}\log(1 + r_D(A)) \leqslant \log\left(1 + \frac{1}{2}r_D(A)\right) \leqslant j_D(A) \leqslant \log(1 + r_D(A)).$$

The first one follows from

$$\log(1 + t) \leqslant \log\left(1 + \frac{t}{2}\right)^2.$$

For the second inequality, we have to show that

$$\log\left(1 + \frac{r_D(A)}{2}\right) \leqslant \sup_{x,y \in A} \log\left(1 + \frac{|x - y|}{\min\{d(x), d(y)\}}\right),$$

i.e.,

$$\frac{d(A)}{d(A, \partial D)} \leqslant 2 \sup_{x,y \in A} \frac{|x - y|}{\min\{d(x), d(y)\}}.$$

Now, for a given $\varepsilon > 0$, we choose an $x_0 \in A$ such that $d(x_0) \leqslant d(A, \partial D) + \varepsilon$ and $x_1, y_1 \in A$ such that $|x_1 - y_1| \geqslant d(A) - \varepsilon$. Then

$$\max\{|x_0 - x_1|, |x_0 - y_1|\} \geqslant \frac{|x_1 - y_1|}{2} \geqslant \frac{d(A) - \varepsilon}{2}$$

and further

$$\max\left\{\frac{|x_0 - x_1|}{d(x_0)}, \frac{|x_0 - y_1|}{d(x_0)}\right\} \geqslant \frac{d(A) - \varepsilon}{2\left(d(A, \partial D) + \varepsilon\right)}$$

because $\varepsilon > 0$ is arbitrary, it follows that

$$\sup_{x,y \in A} \frac{|x - y|}{\min\{d(x), d(y)\}} \geqslant \sup_{x \in A} \frac{|x - x_0|}{\min\{d(x_0), d(x)\}} \geqslant \sup_{x \in A} \frac{|x_0 - x|}{d(x_0)} \geqslant \frac{d(A)}{2\,d(A, \partial D)}.$$

The remaining inequality reduces to

$$\frac{|x - y|}{\min\{d(x), d(y)\}} \leqslant \frac{d(A)}{d(A, \partial D)}, \qquad \text{for all } x, y \in A,$$

which is obviously true because $d(A) \geqslant |x - y|$ and $d(A, \partial D) \leqslant d(x), d(y)$.

For the case of the unit ball $\mathbb{B}^n$ one can develop the properties of the hyperbolic metric $\rho_{\mathbb{B}^n}$ in the same way as for $\rho_{\mathbb{H}^n}$. For basic results we refer to [24] and [158, pp.19–32].

Lemma 3.13 ([10, Lemma 7.56] and [158, Lemma 2.41(2), p. 29])

1. $j_{\mathbb{B}^n}(x, y) \leqslant \rho_{\mathbb{B}^n}(x, y) \leqslant 2\, j_{\mathbb{B}^n}(x, y)$ for $x, y \in \mathbb{B}^n$,
2. $j_{\mathbb{H}^n}(x, y) \leqslant \rho_{\mathbb{H}^n}(x, y) \leqslant 2\, j_{\mathbb{H}^n}(x, y)$ for $x, y \in \mathbb{H}^n$.

3.5 Quasihyperbolic Metric k_D

If ρ is a continuous function with $\rho(x) > 0$ for $x \in D$ and if γ is a rectifiable curve in D, then we can define

$$l_\rho(\gamma) = \int_\gamma \rho \, ds.$$

The Euclidean length of a curve γ is denoted by $l(\gamma)$.

Also, for $x_1, x_2 \in D$ we define

$$d_\rho(x, y) = \inf l_\rho(\gamma), \tag{3.16}$$

where the infimum is taken over all rectifiable curves from x_1 to x_2.

It is easy to show that d_ρ is a metric in D.

A special case of this metric is the hyperbolic metric in $\mathbb{H}^n$ ($\rho(x) = 1/x_n$) and in $\mathbb{B}^n$ ($\rho(x) = 2/(1 - |x|^2)$).

Now we take each proper domain $D \subset \mathbb{R}^n$ and set $\rho(x) = 1/(d(x, \partial D))$.

The corresponding metric, denoted by k_D, is called quasihyperbolic metric in D. Since,

$$\rho(\varphi(x)) = \frac{1}{d(\varphi(x), \partial (\varphi D))} = \frac{1}{d(x, \partial D)} = \rho(x),$$

for Euclidean isometry φ, we have that

$$k_{D'}(x', y') = k_D(x, y), \qquad \text{where } D' = \varphi(D), \ x' = \varphi(x), \ y' = \varphi(y).$$

Moreover, we have that both metrics j_D and k_D are invariant under similarity transformations.

Using a compactness argument (the Helly selection principle), one proves the existence of geodesics:

Lemma 3.14 ([53, Lemma 1]) *For each pair of points $x_1, x_2 \in D$ there exists a quasihyperbolic geodesic γ with x_1, x_2 as its end points.*

For all $x_3 \in \gamma$, we have

$$k_D(x_1, x_2) = k_D(x_1, x_3) + k_D(x_3, x_2).$$

We give a simple proof of the inequality

$$k_D(x, y) \geqslant j_D(x, y)$$

for $x, y \in D$.

Lemma 3.15 ([54, Lemma 2.1]) *For* $x, y \in D$

$$k_D(x, y) \geq \log \left(1 + \frac{|x - y|}{\min\{d(x), d(y)\}} \right) \geq j_D(x, y).$$

Proof We can assume $0 < d(x) \leq d(y)$. Choose a rectifiable arc $\gamma : [0, s] \to D$ from x to y, parametrized by arc length:

$$\gamma(0) = x, \qquad \gamma(s) = y.$$

Obviously, $s \geq |x - y|$. For each $0 \leq t \leq s$, we have the key observation,

$$d(\gamma(t)) \leq d(x) + t,$$

so,

$$l_\rho(\gamma) \geq \int_0^s \frac{dt}{d(x) + t} = \log \frac{d(x) + s}{d(x)} \geq \log \frac{d(x) + |x - y|}{d(x)} = j_D(x, y).$$

$\square$

Lemma 3.16 ([158, Lemma 3.7, p. 34])

1. *If* $x \in D$, $y \in B_x = B^n(x, d(x))$, *then*

$$k_D(x, y) \leq \log \left(1 + \frac{|x - y|}{d(x) - |x - y|} \right).$$

2. *If* $s \in (0, 1)$ *and* $|x - y| \leq s\, d(x)$, *then*

$$k_D(x, y) \leq \frac{1}{1 - s} j_D(x, y).$$

Proof

1. Select $z \in \partial B_x$ such that $y \in [x, z]$, see Fig. 3.6.
   Because $[x, y] \in \Gamma_{xy}$, we obtain

$$k_D(x, y) \leq k_{B_x}(x, y) \leq \int_{[x,y]} \frac{|dw|}{d(w)} \leq \int_{[x,y]} \frac{|dw|}{|w - z|} = \int_{d(x)-|x-y|}^{d(x)} \frac{dt}{t}$$

$$= \log \frac{d(x)}{d(x) - |x - y|} = \log \left(1 + \frac{|x - y|}{d(x) - |x - y|} \right)$$

$$\left(= j_{\mathbb{R}^n \setminus \{z\}}(x, y) \right).$$

Fig. 3.6 Lemma 3.16

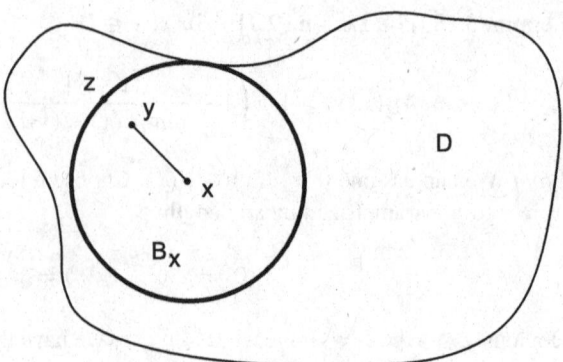

2. For the proof of this part we apply the previous part, the Bernoulli inequality, and the definition of j_D, to obtain

$$k_D(x, y) \leqslant \log\left(1 + \frac{|x - y|}{(1 - s)\,d(x)}\right)$$

$$\leqslant \frac{1}{1 - s}\,\log\left(1 + \frac{|x - y|}{d(x)}\right) \leqslant \frac{1}{1 - s}\,j_D(x, y),$$

as desired.

$\square$

We note that, in general, there is no constant C, such that for all $x, y \in D$

$$k_D(x, y) \leqslant C \cdot j_D(x, y).$$

A domain $D \subset \mathbb{R}^n$, for which the above inequality holds for some $C \geq 1$, is called a uniform domain. Furthermore, the best possible number

$$C_G := \inf\{C \geq 1 : k_G \leq C\,j_G\}$$

is called the uniformity constant of G.

An example of domain which is not uniform is

$$\{x \in \mathbb{R}^2 : |x| < 1\} \setminus \{t\,e_1 : t \geq 0\}.$$

This wide and useful class of uniform domains was introduced by O. Martio and J. Sarvas (see [112]). There are other equivalent definitions for uniform domains; see, for example, the book [52, Definition 3.5.1, p. 40] of F. W. Gehring and K. Hag.

Uniformity of different domains and respective uniformity constants has been studied by H. Linden (see [97]).

3.6 Other Hyperbolic Type Metrics

One can define hyperbolic type metrics in several ways. For example, these metrics can be defined as weighted metrics such as the quasihyperbolic metric or by a formula involving Euclidean distances.

Let us introduce first the *Seittenranta metric* δ_G [141]. For an open set $G \subset \mathbb{R}^n$ with card $\partial G \geqslant 2$ we set

$$m_G(x, y) = \sup_{a,b \in \partial G} |a, x, b, y|$$

and

$$\delta_G(x, y) = \log(1 + m_G(x, y))$$

for all $x, y \in G$.

The case of an unbounded domain $G \subset \mathbb{R}^n, \infty \in \partial G$, should now be considered. Note that if a or b in the supremum is equal to the infinity, then we get exactly the j_G metric. This implies that we always have $j_G \leqslant \delta_G$.

Let us also introduce the *Apollonian metric* considered by Beardon [25], (see also [10, 7.28 (2)]) defined in open proper subsets $G \subset \mathbb{R}^n$ as follows:

$$\alpha_G(x, y) = \sup_{a,b \in \partial G} \log |a, x, y, b| \quad \text{for all } x, y \in G.$$

This formula defines a metric if and only if $\mathbb{R}^n \setminus G$ is not a proper subset of an $(n-1)$-dimensional sphere in $\mathbb{R}^n$.

For a domain $G \subset \mathbb{R}^n$ and $x, y \in G$, the *triangular ratio metric* s_G is defined as follows:

$$s_G(x, y) = \sup_{z \in \partial G} \frac{|x - y|}{|x - z| + |z - y|} \in [0, 1].$$

This metric was introduced by P. Hästö [59] and studied in [58].

Another group of hyperbolic type metrics may again be classified by the number of boundary points used in their definition. So for instance, the j metric is a one-point metric, while the Apollonian metric is a two-point metric.

Also, it is natural to introduce conformal invariants $\lambda_G(x, y)$ and $\mu_G(x, y)$ defined for a domain $G \subset \mathbb{R}^n$ and $x, y \in G$. A basic fact is that $\lambda_G(x, y)^{1/(1-n)}$ and $\mu_G(x, y)$ are metrics [10, Remark 16.18 (2), p. 320].

If G is a proper subdomain of $\overline{\mathbb{R}^n}$, then for $x, y \in G$ with $x \neq y$ we define

$$\lambda_G(x, y) = \inf_{C_x, C_y} M(\Delta(C_x, C_y; G)) \tag{3.17}$$

where $C_z = \gamma_z[0, 1)$ and $\gamma_z : [0, 1) \longrightarrow G$ is a curve such that $\gamma_z(0) = z$ and $\gamma_z(t) \to \partial G$ when $t \to 1$, $z = x, y$. This conformal invariant was introduced by J. Ferrand (see [159]).

If G is a proper subdomain of $\overline{\mathbb{R}^n}$, then for $x, y \in G$, we define

$$\mu_G(x, y) = \inf_{C_{xy}} M(\Delta(C_{xy}, \partial G; G)), \tag{3.18}$$

where the infimum is taken over all continua C_{xy} such that $C_{xy} = \gamma[0, 1]$ and γ is a curve with $\gamma(0) = x$ and $\gamma(1) = y$. In the case $G = \mathbb{B}^n$ the function $\mu_{\mathbb{B}^n}(x, y)$ is the extremal quantity of H. Grötzsch (see [159]).

For further relations between other hyperbolic type metrics the reader is referred to [25, 61, 101, 128] and [141].

3.7 Quasiconformal Mappings and k_D and j_D Metrics

Gehring and Osgood [53] proved a quasi-invariance property of the quasihyperbolic metric. To prove this property, we need the following lemma, obtained by combining (2.16) with the distortion theorem 2.48.

Lemma 3.17 ([53, Lemma 2]) *For $n \geq 2$ there exists a constant, a, depending only on n with the following property. If f is a K-quasiconformal mapping of D onto D', then*

$$\frac{|f(x_1) - f(x_2)|}{d(f(x_1), \partial D')} \leqslant a \left(\frac{|x_1 - x_2|}{d(x_1, \partial D)}\right)^\alpha, \qquad \alpha = K^{1/(1-n)} \tag{3.19}$$

for all $x_1, x_2 \in D$ with

$$\frac{|x_1 - x_2|}{d(x_1, \partial D)} \leqslant a^{-1/\alpha}.$$

Theorem 3.18 ([53, Theorem 3]) *For $n \geqslant 2$, $K \geqslant 1$, there exists a constant c depending only on n and K with the following property. If f is a K-quasiconformal mapping of D onto D', then*

$$k_{D'}(f(x_1), f(x_2)) \leqslant c \cdot \max\{k_D(x_1, x_2), k_D(x_1, x_2)^\alpha\}, \qquad \alpha = K^{1/(1-n)}, \tag{3.20}$$

for all $x_1, x_2 \in D$.

Proof We split the proof into two cases.

Case A Suppose that

$$\frac{|x_1 - x_2|}{d(x_1, \partial D)} \leqslant (2a)^{-1/\alpha} < 1. \tag{3.21}$$

By the previous lemma

$$\frac{|f(x_1) - f(x_2)|}{d(f(x_1), \partial D')} \le a \left(\frac{|x_1 - x_2|}{d(x_1, \partial D)} \right)^{\alpha} \le \frac{1}{2} \tag{3.22}$$

and

$$d(y, \partial D') \ge \frac{1}{2} d(f(x_1), \partial D') \tag{3.23}$$

for all y on the segment joining $f(x_1)$ and $f(x_2)$.

Hence,

$$k_{D'}(f(x_1), f(x_2)) \le \frac{2|f(x_1) - f(x_2)|}{d(f(x_1), \partial D')} \le 1. \tag{3.24}$$

Next,

$$k_D(x_1, x_2) \ge \log \left(\frac{|x_1 - x_2|}{d(x_1, \partial D)} + 1 \right) \ge \frac{1}{2} \frac{|x_1 - x_2|}{d(x_1, \partial D)}. \tag{3.25}$$

The first inequality follows from Lemma 3.15. The second inequality follows from the following simple consequence of the concavity of the logarithm function,

$$\log(1 + t) \ge t \log 2 \ge \frac{t}{2}, \qquad \text{for } 0 \le t \le 1,$$

applied with $t = \frac{|x_1 - x_2|}{d(x_1, \partial D)}$. From (3.22) and (3.25) we obtain

$$k_{D'}(f(x_1, f(x_2)) \le 4 a \, k_D(x_1, x_2)^{\alpha}. \tag{3.26}$$

Case B Now suppose that (3.21) is not true. Then join x_1, x_2 by a quasihyperbolic geodesic curve γ and pick $y_1 = x_1, \ldots, y_{m+1} = x_2$ and so that

$$\frac{|y_j - y_{j+1}|}{d(y_j, \partial D)} = (2 a)^{1-\alpha}, \qquad \frac{|y_m - y_{m+1}|}{d(y_m, \partial D)} \le (2 a)^{-1/\alpha}$$

for $j = 1, \ldots, m - 1$. Then

$$k_{D'}(f(x_1), f(x_2)) \le \sum_{j=1}^{m} k_{D'}(f(y_j), f(y_{j+1})) \le m,$$

because each term is ≤ 1 by (3.24).

Since the points y_j lie on geodesics, we have by additivity on geodesics (Lemma 3.14)

$$k_D(x_1, x_2) = \sum_{j=1}^{m} k_D(y_j, y_{j+1}) \ge \frac{m-1}{2} (2 a)^{-1/\alpha}$$

by (3.25). Thus

$$k_{D'}(f(x_1), f(x_2)) \leqslant 4(2a)^{1/\alpha} k_D(x_1, x_2) \tag{3.27}$$

since $m \geqslant 2$. Inequality (3.20) then follows from (3.26) and (3.27) with $c = 4(2a)^{1/\alpha}$.

□

It was proved in [83, Theorem 1.4] that the quasihyperbolic metric is not invariant under Möbius transformations of the unit ball onto itself, and hence the constant c in Theorem 3.18 cannot be asymptotically sharp when $K \to 1$. In other words, $c \not\to 1$ when $K \to 1$. This is studied also in [9] where it is shown that $c \to \infty$ when $K \to \infty$ and moreover that c can be chosen independently on n. This means that there is a dimension-free variant of Theorem 3.18 [158, Cor. 12.19, 12.20].

In the Theorem 3.19 below, for the special domain $G = \mathbb{R}^n \setminus \{0\}$, a quasi-invariance property of the quasihyperbolic metric with the asymptotically sharp constants is established. We remark that for this domain Martin and Osgood [109] proved that

$$k_D(x, y) = \sqrt{\log^2 \frac{|x|}{|y|} + \left(2 \arcsin \left(\frac{1}{2} \left(\left| \frac{x}{|x|} - \frac{y}{|y|} \right| \right) \right) \right)^2}, \tag{3.28}$$

for all $x, y \in D$.

Theorem 3.19 ([84]) *For given $K \in (1, 2]$ and $n \geq 2$ there exists a constant $\omega(K, n)$ such that if $G = \mathbb{R}^n \setminus \{0\}$ and $f : \mathbb{R}^n \to \mathbb{R}^n$ is a K-quasiconformal mapping with $f(0) = 0$, then for all $x, y \in G$*

$$k_G(f(x), f(y)) \leq \omega(K, n) \max\{k_G(x, y)^\alpha, k_G(x, y)\}$$

where $\alpha = K^{1/(1-n)}$ and $\omega(K, n) \to 1$ when $K \to 1$.

We note that Agard and Gehring have studied the angle distortion under quasiconformal mappings of the plane [2]. Motivated by their work we state the following corollary of Theorem 3.19.

Corollary 3.20 ([84]) *Suppose that under the hypotheses of Theorem 3.19, $x, y \in S^{n-1}$ and $f(x), f(y) \in S^{n-1}$. Let ϕ and ψ be the angles between the segments $[0, x], [0, y]$ and $[0, f(x)], [0, f(y)]$ respectively. Then*

$$\psi \leq \omega(K, n) \max\{\phi^\alpha, \phi\}.$$

Proof The proof follows easily from the Martin–Osgood formula (3.28). Since $x, y, f(x), f(y) \in S^{n-1}$ holds $|x| = |y| = |f(x)| = |f(y)| = 1$ and $\log(|x|/|y|) = 0$. Then

$$k_G(x, y) = 2 \arcsin(|x - y|/2) = \phi$$

and

$$k_G(f(x), f(y)) = 2 \arcsin(|f(x) - f(y)|/2) = \psi.$$

$\square$

The proof of the following Theorem 3.21 [84] relies on two ingredients: a sharp version of the Schwarz lemma for quasiconformal mappings from [9], and a sharp bound for the linear dilatation from [159]. We note that Theorem 3.21 represents an analogue of Theorem 3.19 for the distance ratio metric.

Theorem 3.21 ([84]) *For given* $K \in (1, 2]$ *and* $n \geq 2$ *there exist a constant* $c(K)$ *such that if* $G = \mathbb{R}^n \setminus \{0\}$ *and* $f : \mathbb{R}^n \longrightarrow \mathbb{R}^n$ *is a* K*-quasiconformal mapping with* $f(0) = 0$, *then for all* $x, y \in G$

$$j_G(f(x), f(y)) \leq c(K) \max\{j_G(x, y)^\alpha, j_G(x, y)\},$$

where $\alpha = K^{1/(1-n)}$, *and* $c(K) \to 1$ *as* $K \to 1$.

3.8 Quasiconformal Mappings with Identity Boundary Values

The paper [84] considers the standard normalization which requires that the mapping keeps two points fixed and proves a stability result for dimensions $n \geq 2$, which is a counterpart of O. Teichmüller's result in the case $n = 2$. We recall that the well-known Verschiebungssatz of Teichmüller [152] was a surprisingly new phenomenon of quasiconformality which later became a useful tool in many applications. It concerns the minimal dilatation of quasiconformal automorphisms of the disk, which have identical boundary values and move one prescribed inner point onto another. The extremal dilatation is given in terms of hyperbolic distance between these points (with the explicit description of the extremal map) and so the result is extended to each hyperbolic domain on the Riemann sphere. Namely, for a domain $G \subset \mathbb{R}^n, n \geqslant 2$, let

$$Id(\partial G) = \{f : \overline{\mathbb{R}^n} \to \overline{\mathbb{R}^n} \text{ homeomorphism} : f(x) = x, \quad \forall x \in \overline{\mathbb{R}^n} \setminus G\}.$$

Recall that here $\overline{\mathbb{R}^n}$ stands for the Möbius space $\mathbb{R}^n \cup \{\infty\}$. We shall always assume that $card\{\overline{\mathbb{R}^n} \setminus G\} \geq 3$. If $K \geqslant 1$, then the class of K-quasiconformal maps in $Id(\partial G)$ is denoted by $Id_K(\partial G)$.

O. Teichmüller proved the following theorem with a sharp bound for $K(f)$.

Theorem 3.22 *Let* $G = \mathbb{R}^2 \setminus \{0, 1\}$, $a, b \in G$. *Then there exists* $f \in Id_K(\partial G)$ *with* $f(a) = b$ *iff*

$$\log(K(f)) \geqslant s_G(a, b),$$

where $s_G(a, b)$ *is the hyperbolic metric of* G.

The search for a multidimensional analog of this result and questions related to this have been investigated for quite some time, but so far only special results have been obtained. For example, in [104] the author and M. Vuorinen have studied the problem of multidimensional version of Teichmüller's theorem. The main result of [104] is an upper bound for the hyperbolic distance $\rho(f(x), x)$ for admissible K-quasiconformal automorphisms of the unit ball $\mathbb{B}^n$ in $\mathbb{R}^n$. This bound holds for all points of $\mathbb{B}^n$.

Lemma 3.23 ([104]) *For $x, y \in \mathbb{B}^n$ let $t = \sqrt{(1 - |x|^2)(1 - |y|^2)}$. Then for $x, y \in \mathbb{B}^n$*

$$\text{th}^2 \frac{\rho_{\mathbb{B}^n}(x, y)}{2} = \frac{|x - y|^2}{|x - y|^2 + t^2}, \tag{3.29}$$

$$|x - y| \leqslant 2\,\text{th}\frac{\rho_{\mathbb{B}^n}(x, y)}{4} = \frac{2|x - y|}{\sqrt{|x - y|^2 + t^2} + t}, \tag{3.30}$$

where equality holds for $x = -y$.

Theorem 3.24 ([104]) *If $f \in Id_K(\partial \mathbb{B}n)$, then for all $x \in \mathbb{B}^n$*

$$\rho_{\mathbb{B}^n}(f(x), x) \leqslant \log \frac{1 - a}{a}, \quad a = \varphi_{1/K,n}(1/\sqrt{2})^2, \tag{3.31}$$

where $\varphi_{K,n}$ is as in (2.7).

Proof Fix $x \in \mathbb{B}^n$ and let T_x denote a Möbius transformation of $\overline{\mathbb{R}^n}$ with $T_x(\mathbb{B}^n) = \mathbb{B}^n$ and $T_x(x) = 0$. Define $g : \mathbb{R}^n \longrightarrow \mathbb{R}^n$ by setting $g(z) = T_x \circ f \circ T_x^{-1}(z)$ for $z \in \mathbb{B}^n$ and $g(z) = z$ for $z \in \mathbb{R}^n \setminus \mathbb{B}^n$. Then $g \in Id_K(\partial \mathbb{B}^n)$ with $g(0) = T_x(f(x))$ (Fig. 3.7). By the invariance of $\rho_{\mathbb{B}^n}$ under the group $\mathscr{GM}(\mathbb{B}^n)$ of Möbius selfautomorphisms of $\mathbb{B}^n$, we see that for $x \in \mathbb{B}^n$,

$$\rho_{\mathbb{B}^n}(f(x), x) = \rho_{\mathbb{B}^n}(T_x(f(x)), T_x(x)) = \rho_{\mathbb{B}^n}(g(0), 0). \tag{3.32}$$

Choose $z \in \partial \mathbb{B}^n$ such that $g(0) \in [0, z] = \{tz : 0 \leqslant t \leqslant 1\}$. Let $E' = \{-sz : s \geqslant 1\}$, $\Gamma' = \Delta([g(0), z], E'; \mathbb{R}^n)$ and $\Gamma = \Delta(g^{-1}[g(0), z], g^{-1}E'; \mathbb{R}^n)$.

The spherical symmetrization with center at 0 yields by [10, Theorem 8.44]

$$M(\Gamma) \geqslant \tau_n(1) \quad (= 2^{1-n}\gamma_n(\sqrt{2}))$$

because $g(x) = x$ for $x \in \mathbb{R}^n \setminus \mathbb{B}^n$. Next, we see by the choice of Γ' that

$$M(\Gamma') = \tau_n\left(\frac{1 + |g(0)|}{1 - |g(0)|}\right).$$

Fig. 3.7 The mapping g with identity boundary values

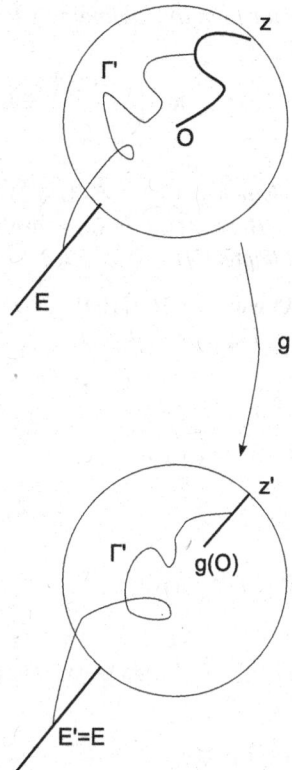

By K-quasiconformality we have $M(\Gamma) \leqslant K\, M(\Gamma')$ implying

$$\exp(\rho_{\mathbb{B}^n}(0, g(0))) = \frac{1 + |g(0)|}{1 - |g(0)|} \leqslant \tau_n^{-1}(\tau_n(1)/K) = \frac{1-a}{a}. \qquad (3.33)$$

The last equality follows from (2.11). Finally, (3.32) and (3.33) complete the proof.
□

Lemma 3.25 ([104]) *If* $a = \varphi_{1/K,n}(1/\sqrt{2})^2$ *is as in Theorem 3.24, then for* $M > 1$ *and* $\beta \in [1, M]$,

$$\log\left(\frac{1-a}{a}\right) \leq \log(\lambda_n^{2(\beta-1)} 2^\beta - 1) \leq V(n)(\beta - 1), \qquad (3.34)$$

where $V(n) = (2\log(2\lambda_n^2))(2\lambda_n^2)^{M-1}$. *Furthermore, for* $K \in [1, 17]$,

$$\log\left(\frac{1-a}{a}\right) \leqslant (K-1)(4 + 6\log 2) < 9(K-1), \qquad (3.35)$$

with equality holding only for K = 1. For n = 2,

$$\log\left(\frac{1-a}{a}\right) = \log\left(\frac{\varphi_{K,2}(1/\sqrt{2})^2}{\varphi_{1/K,2}(1/\sqrt{2})^2}\right) \leq b(K-1), \tag{3.36}$$

where $b = (4/\pi)\mathcal{K}(1/\sqrt{2})^2 \leq 4.38$.

Here $\mathcal{K}(r)$ is Legendre's complete elliptic integral of the first kind (see [10, Chapter 3]).

Theorem 3.26 ([104]) *If $f \in Id_K(\partial\mathbb{B}^n)$, then for all $x \in \mathbb{B}^n, n \geq 2$, and $K \in [1, 17]$*

$$|f(x) - x| \leq \frac{9}{2}(K-1). \tag{3.37}$$

For $n = 2$ we have

$$|f(x) - x| \leq \frac{b}{2}(K-1), \quad b \leq 4.38. \tag{3.38}$$

Proof We have

$$|f(x) - x| \leq 2\,\text{th}\left(\frac{\rho_{\mathbb{B}^n}(f(x), x)}{4}\right) \leq 2\,\text{th}\left(\frac{\log\left(\frac{1-a}{a}\right)}{4}\right)$$

$$\leq 2\,\text{th}\left(\frac{(K-1)(4+6\log 2)}{4}\right)$$

$$\leq (K-1)(2+3\log 2) \leq \frac{9}{2}(K-1).$$

The first inequality follows from (3.30), the second one from Theorem 3.24, the third one from Lemma 3.25, and the last one from the inequality $\text{th}(t) \leq t$ for $t \geq 0$.

For $n = 2$ we use the same first two steps and the planar case of Lemma 3.25 to derive the inequality

$$|f(x) - x| \leq \frac{b}{2}(K-1).$$

$\square$

We next consider the class $Id_K(\partial Z)$ for the case when the domain Z is an infinite cylinder.

Theorem 3.27 ([98]) *Let $Z = \{(x, t) \in \mathbb{R}^n : |x| < 1, t \in \mathbb{R}\}, f \in Id_K(\partial Z)$. Then $k_Z(0, f(0)) \leq c(K)$ where $c(K) \to 0$ when $K \to 1$.*

Proof Let $f(0) = (y, t), E' = [w, f(0)], F' = \{\overline{w} + s(y, 0) : s \leq 0\}$, where $w = (y/|y|, t), \overline{w} = (-y/|y|, t)$ (Fig. 3.8). Then E' and F' are the complementary components of a Teichmüller ring and writing $\Gamma' = \Delta(E', F'; \mathbb{R}^n)$, we have

Fig. 3.8 Maps of cylinder

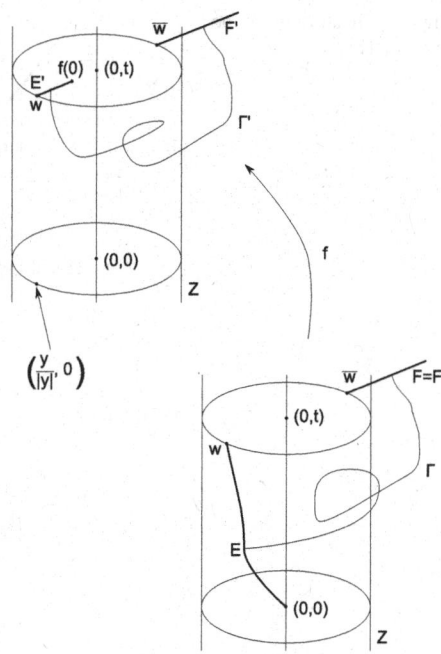

$$M(\Gamma') \leqslant \tau_n \left(\frac{1+|y|}{1-|y|} \right).$$

The modulus of the family $\Gamma = \Delta(E, F; \mathbb{R}^n)$, $E = f^{-1}E'$, $F = f^{-1}F'$ can be estimated by use of spherical symmetrization with the center at 0. Note that $E = E'$ because $E' \subset \mathbb{R}^n \setminus Z$ and $f \in Id_K(\partial Z)$. By [159, 7.34], we have

$$M(\Gamma) \geqslant \tau_n(1).$$

By K-quasiconformality $M(\Gamma) \leqslant K\, M(\Gamma')$ implying

$$\exp(\rho_{B^{n-1}}(0, y)) = \frac{1+|y|}{1-|y|} \leqslant \tau_n^{-1}\left(\frac{\tau_n(1)}{K} \right).$$

Next we shall estimate t. First fix z in $\{w \in \partial Z : w_n = 0\}$ such that $|f(0) - z|$ is maximal. Then choose a point w on the line through $f(0)$ and z such that $|z-w| = 1$ and $[z, w] \subset \mathbb{R}^n \setminus Z$. Let $E' = [z, w]$ and $F' = \{f(0) + t(f(0) - z) : t \geqslant 0\}$. Then E' and F' are the complementary components of a Teichmüller ring and writing $\Delta' = \Delta(E', F'; \mathbb{R}^n)$ (Fig. 3.9), we have

$$M(\Delta') = \tau_n(|f(0) - z|).$$

Fig. 3.9 Illustration for the
proof of Theorem 3.27

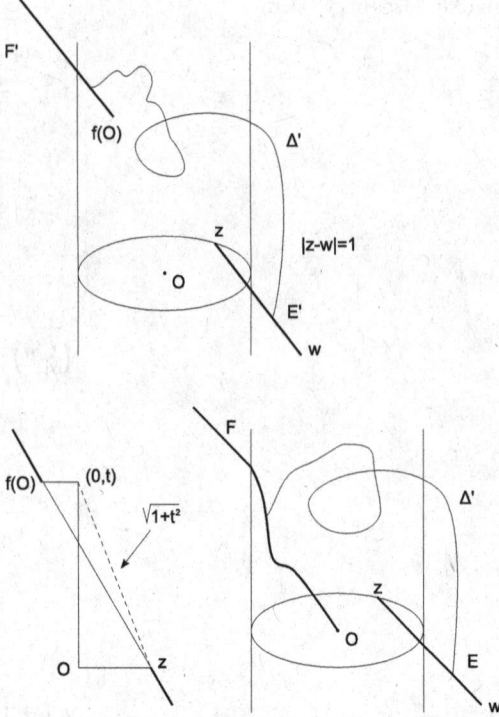

Observing that $E' = f^{-1}E'$ (because $f \in Id_K(\partial Z)$) and carrying out a spherical symmetrization with center at z we see that if $E = f^{-1}E'$, $F = f^{-1}F'$ then

$$M(\Delta) \geqslant \tau_n(1), \quad \Delta = \Delta(E, F; \mathbb{R}^n).$$

By K-quasiconformality we have

$$1 + t^2 \leqslant |f(0) - z|^2 \leqslant \tau_n^{-1}\left(\frac{\tau_n(1)}{K}\right)^2.$$

The triangle inequality for k_Z yields

$$
\begin{aligned}
k_Z(0, f(0)) &\leqslant k_Z(0, (0, t)) + k_Z((0, t), (y, t)) \\
&= t + k_{B^{n-1}}(0, y) \leqslant |t| + 2\,\rho_{B^{n-1}}(0, y) \\
&\leqslant \sqrt{\tau_n^{-1}\left(\frac{\tau_n(1)}{K}\right)^2 - 1} + 2\log\left(\tau_n^{-1}\left(\frac{\tau_n(1)}{K}\right)\right) \\
&\leqslant \sqrt{e^{18(K-1)} - 1} + 18(K - 1).
\end{aligned}
$$

The last inequality follows from (2.11) and Lemma 3.25. $\square$

We finish this chapter by remarking that recently several other authors have studied extensions, ramifications, and generalizations of Teichmüller's problem. For example, the case of this problem for the unit ball in $\mathbb{R}^n$ is considered in [160, 163] and the case of Riemann surfaces in [27].

Chapter 4
Distance Ratio Metric

The basic distortion results about quasiconformal mappings such as the Schwarz lemma and the Gehring–Osgood theorem say that these mappings are Hölder continuous with respect to the hyperbolic and the quasihyperbolic metric respectively. In this chapter we analyze the modulus of continuity in the case of the distance ratio metric. The natural question is to find Lipschitz constants for this metric under Möbius transformations or arbitrary holomorphic mappings. The domains we work with here are the unit ball, the punctured ball, and the upper half space.

Recall that $a^* = \frac{a}{|a|^2}$ for $a \in \mathbb{R}^n \setminus \{0\}$ and that $0^* = \infty$, $\infty^* = 0$. For fixed $a \in \mathbb{B}^n \setminus \{0\}$, let

$$\sigma_a(x) = a^* + s^2(x - a^*)^*, \quad s^2 = |a|^{-2} - 1 \tag{4.1}$$

be an inversion in the sphere $S^{n-1}(a^*, s)$ orthogonal to S^{n-1} see (3.1) and (3.2). Then $\sigma_a(a) = 0, \sigma_a(0) = a, \sigma_a(a^*) = \infty$ and

$$|\sigma_a(x) - \sigma_a(y)| = \frac{s^2|x - y|}{|x - a^*||y - a^*|} \tag{4.2}$$

(see Lemma 3.1).

Lemma 4.1 ([24, Theorem 3.5.1, p. 40]) *Let f be a Möbius transformation and $f(\mathbb{B}^n) = \mathbb{B}^n$. Then*

$$f(x) = (\sigma x)A,$$

where σ is an inversion in some sphere orthogonal to S^{n-1} and A is an orthogonal matrix.

Since the j-metric is invariant under orthogonal transformations, by Lemma 4.1, for $x, y, a \in \mathbb{B}^n$, we have

© Springer Nature Switzerland AG 2019
V. Todorčević, *Harmonic Quasiconformal Mappings and Hyperbolic Type Metrics*,
https://doi.org/10.1007/978-3-030-22591-9_4

$$j_{\mathbb{B}^n}(f(x), f(y)) = j_{\mathbb{B}^n}(\sigma_a(x), \sigma_a(y)),$$

where $\sigma_a(x)$ is defined as above and f is a Möbius transformation.

For given domains $D, D' \subset \mathbb{R}^n$ and an open continuous mapping $f : D \to D'$ with $fD \subset D'$, let us consider the following condition: There exists a constant $C \geq 1$ such that for all $x, y \in D$ we have

$$j_{D'}(f(x), f(y)) \leq C j_D(x, y), \tag{4.3}$$

or, equivalently, that the mapping

$$f : (D, j_D) \to (D', j_{D'})$$

between metric spaces is Lipschitz continuous with the Lipschitz constant C.

The process of determining the best possible j-Lip constants is characterized by a series of subtle inequalities that are of interest to study. It is well known that the hyperbolic metric in the unit ball or half space is Möbius invariant. This is not true for the distance ratio metric j_G, and so it is natural to ask what is the Lipschitz constant for this metric under conformal mappings or Möbius transformations in higher dimension. F. W. Gehring and B. G. Osgood proved that this metric is not changed by more than a factor of 2 under Möbius transformations (see [53, proof of Theorem 4]).

Theorem 4.2 *If D and D' are proper subdomains of $\mathbb{R}^n$ and if $f : \overline{\mathbb{R}}^n \to \overline{\mathbb{R}}^n$ is a Möbius transformation with $f(D) = D'$, then for all $x, y \in D$*

$$\frac{1}{2} j_D(x, y) \leq j_{D'}(f(x), f(y)) \leq 2 j_D(x, y).$$

The purpose of this chapter examines the challenging task of improving this estimation for some particular domains of $\mathbb{R}^n$. Our presentation here is based on the papers [146, 147] and [143].

4.1 Refinements of the Gehring–Osgood Result

One of the refinements of the Gehring–Osgood result is given by S. Simić and M. Vuorinen [146] and is of a two-fold nature. First, the constant 2 can be essentially improved in the cases of the unit ball and the punctured unit ball in $\mathbb{R}^n$ with the best possible constants provided. On the other hand, it happens that Theorem 4.2 is valid on the half-plane or the unit disk not only for Möbius transformation but also for arbitrary holomorphic mappings. Proofs of those facts are based on the famous Schwarz–Pick Lemma.

The following theorem conjectured in [83] and proved by S. Simić, M. Vuorinen, and G. Wang [147] yields a sharp form of Theorem 4.2 for Möbius automorphisms of the unit ball.

Theorem 4.3 ([147]) *A Möbius transformation* $f : \mathbb{B}^n \to \mathbb{B}^n = f(\mathbb{B}^n),\ f(0) = a \in \mathbb{B}^n,\ satisfies$

$$\frac{1}{1+|a|} j_{\mathbb{B}^n}(x, y) \le j_{\mathbb{B}^n}(f(x), f(y)) \le (1+|a|) j_{\mathbb{B}^n}(x, y)$$

for all $x, y \in \mathbb{B}^n$. *The constants are best possible.*

In order to prove this result, we introduce a series of lemmas.

Lemma 4.4 ([158, Exercise 2.52 (2), p. 32]) *For the hyperbolic metric there holds*

$$\frac{|x| - |y|}{1 - |x|\,|y|} \le \mathrm{th}\frac{1}{2}\rho(x, y) \le \frac{|x| + |y|}{1 + |x|\,|y|}. \tag{4.4}$$

Proof Starting with the identity

$$\mathrm{th}\frac{1}{2}\rho(x, y) = \frac{|x - y|}{\sqrt{|x - y|^2 + (1 - |x|^2)(1 - |y|^2)}} \tag{4.5}$$

it suffices to prove

$$\frac{|x| - |y|}{1 - |x|\,|y|} \le \frac{|x - y|}{\sqrt{|x - y|^2 + (1 - |x|^2)(1 - |y|^2)}} \le \frac{|x| + |y|}{1 + |x|\,|y|}. \tag{4.6}$$

If $|x| < |y|$, then the first inequality holds because the left-hand side is a negative real. Otherwise, the double inequality (4.6) is about nonnegative reals. For this reason, by squaring this double inequality we obtain the following equivalent inequality:

$$\frac{|x|^2 + |y|^2 - 2|x|\,|y|}{1 + |x|^2\,|y|^2 - 2|x|\,|y|} \le \frac{|x|^2 + |y|^2 - 2\langle x, y\rangle}{1 + |x|^2\,|y|^2 - 2\langle x, y\rangle} \le \frac{|x|^2 + |y|^2 + 2|x|\,|y|}{1 + |x|^2\,|y|^2 + 2|x|\,|y|}. \tag{4.7}$$

Recall the Cauchy inequality $|\langle x, y\rangle| \le |x|\,|y|$. Applying it, we get that

$$-2|x|\,|y| \le -2\langle x, y\rangle \le 2|x|\,|y|. \tag{4.8}$$

Because $|x|, |y| < 1$, we have

$$1 + |x|^2\,|y|^2 = (1 - |x|\,|y|)^2 + 2|x|\,|y| > 2|x|\,|y|.$$

It follows that the function

$$\varphi(t) = \frac{|x|^2 + |y|^2 + t}{1 + |x|^2\,|y|^2 + t}$$

is defined for $t \geq -2|x||y|$, and it is increasing because

$$\varphi(t) = 1 - \frac{(1 - |x|^2)(1 - |y|^2)}{1 + |x|^2 |y|^2 + t}.$$

From this and the inequality (4.8), we get (4.7).   □

We shall need the following simple fact about the unit ball.

Lemma 4.5 *Let $a, b \in \mathbb{B}^n$. Then*

1.

$$|a|^2 |b - a^*|^2 - |b - a|^2 = (1 - |a|^2)(1 - |b|^2).$$

2.

$$\frac{||b| - |a||}{1 - |a||b|} \leq \frac{|b - a|}{|a||b - a^*|} \leq \frac{|b| + |a|}{1 + |a||b|}.$$

Proof

1. Using a simple calculation, we get

$$|a|^2 |b - a^*|^2 - |b - a|^2$$

$$= |a|^2 [|b|^2 + \frac{1}{|a|^2} - \frac{2(b \cdot a)}{|a|^2}] - [|b|^2 + |a|^2 - 2(b \cdot a)]$$

$$= 1 + |a|^2 |b|^2 - |a|^2 - |b|^2 = (1 - |a|^2)(1 - |b|^2).$$

2. This can be obtained by Lemma 4.4 directly.

□

We shall need the following *monotone form of l'Hôpital's rule.*

Lemma 4.6 ([10, Theorem 1.25]) *For $-\infty < a < b < \infty$, let $f, g : [a, b] \to \mathbb{R}$ be continuous on $[a, b]$ and differentiable on (a, b), and suppose $g'(x) \neq 0$ on (a, b). If $f'(x)/g'(x)$ is increasing/deceasing on (a, b), then so are*

$$\frac{f(x) - f(a)}{g(x) - g(a)} \quad and \quad \frac{f(x) - f(b)}{g(x) - g(b)}.$$

If $f'(x)/g'(x)$ is strictly monotone, then the monotonicity in the conclusion is also strict.

It should be mentioned that Lemma 4.6 has found numerous applications. See, for example, [43] and the bibliography of [11] for a long list of applications towards various kinds of inequalities.

Lemma 4.7 *Let* $c, d \in (0, 1), \theta \in (0, 1]$. *Then*

1. $f(\theta) \equiv \dfrac{\log\left(1+\frac{2cd\theta}{1-cd}\right)}{\log\left(1+\frac{2d\theta}{1-d}\right)}$ *is increasing. In particular,*

$$\frac{\log\left(1+\frac{2cd\theta}{1-cd}\right)}{\log\left(1+\frac{2d\theta}{1-d}\right)} \leq \frac{\log\left(1+\frac{2cd}{1-cd}\right)}{\log\left(1+\frac{2d}{1-d}\right)}.$$

2. $g(\theta) \equiv \dfrac{\mathrm{arth}(c\theta)}{\mathrm{arth}\theta}$ *is a decreasing function. In particular,*

$$\frac{\mathrm{arth}(c\theta)}{\mathrm{arth}\theta} \leq c.$$

3.

$$\left(1+\frac{2cd\theta}{1-cd}\right)\left(1+\frac{c(1-d)}{1+cd}\right) \leq 1+\frac{c(1-d)+2cd\theta}{1-cd}.$$

Proof

1. Let $f_1(\theta) = \log\left(1+\frac{2cd\theta}{1-cd}\right)$ and $f_2(\theta) = \log\left(1+\frac{2d\theta}{1-d}\right)$. Then we have $f_1(0^+) = f_2(0^+) = 0$ and

$$\frac{f_1'(\theta)}{f_2'(\theta)} = 1 - \frac{1-c}{1-cd+2cd\theta},$$

which is increasing in θ. Hence, the monotonicity of f follows from Lemma 4.6 above. The inequality follows by the monotonicity of f.

2. Let $g_1(\theta) = \mathrm{arth}(c\theta)$ and $g_2(\theta) = \mathrm{arth}\theta$. Then we have $g_1(0^+) = g_2(0^+) = 0$ and

$$\frac{g_1'(\theta)}{g_2'(\theta)} = \frac{1}{c}\left(1 - \frac{1-c^2}{1-c^2\theta^2}\right)$$

which is clearly decreasing in θ. Hence, the monotonicity of g follows from Lemma 4.6. The inequality follows by the monotonicity of g and l'Hôpital's rule.

3. The proof of this is straightforward and left to the reader.

□

We are now in a position to give a short proof of Theorem 4.3.

Proof The conclusion is trivial for $a = 0$, so we only need to consider $a \neq 0$. Since the j-metric is invariant under orthogonal transformations, using Lemma 4.1 for $x, y, a \in \mathbb{B}^n$, we have

$$j_{\mathbb{B}^n}(f(x), f(y)) = j_{\mathbb{B}^n}(\sigma_a(x), \sigma_a(y)),$$

where $\sigma_a(x)$ is an inversion in the sphere $S^{n-1}(a^*, \sqrt{|a|^{-2}-1})$ orthogonal to S^{n-1}. Hence, it suffices to estimate the expression

$$J(x, y; a) \equiv \frac{j_{\mathbb{B}^n}(\sigma_a(x), \sigma_a(y))}{j_{\mathbb{B}^n}(x, y)} = \frac{\log\left(1 + \frac{|\sigma_a(x) - \sigma_a(y)|}{\min\{1-|\sigma_a(x)|, 1-|\sigma_a(y)|\}}\right)}{\log\left(1 + \frac{|x-y|}{\min\{1-|x|, 1-|y|\}}\right)}.$$

Let $r = \max\{|x|, |y|\}$ and suppose $|\sigma_a(x)| \geq |\sigma_a(y)|$. Then by (4.2), we have

$$\min\{1 - |\sigma_a(x)|, 1 - |\sigma_a(y)|\} = 1 - |\sigma_a(x)| = \frac{|a||x - a^*| - |x - a|}{|a||x - a^*|}.$$

Let us first prove the right-hand side of the inequality. By Lemma 4.5, we get

$$j_{\mathbb{B}^n}(\sigma_a(x), \sigma_a(y)) = \log\left(1 + \frac{(1 - |a|^2)|x - y|}{|a||y - a^*|(|a||x - a^*| - |x - a|)}\right)$$

$$= \log\left(1 + \frac{|x - y|(|a||x - a^*| + |x - a|)}{|a||y - a^*|(1 - |x|^2)}\right)$$

$$= \log\left(1 + \frac{|x - y||x - a^*|)}{(1 - |x|^2)|y - a^*|}\left(1 + \frac{|x - a|}{|a||x - a^*|}\right)\right)$$

$$\leq \log\left(1 + \frac{|x - y|}{1 - r^2}\left(1 + \frac{|x - y|}{|y - a^*|}\right)\left(1 + \frac{|x| + |a|}{1 + |a||x|}\right)\right)$$

$$\leq \log\left(1 + \frac{|x - y|}{1 - r}\left(1 + \frac{|a||x - y|}{1 - |a|r}\right)\left(1 + \frac{|a|(1 - r)}{1 + |a|r}\right)\right).$$

Then

$$J(x, y; a) \leq \frac{\log\left(1 + \frac{|x-y|}{1-r}\left(1 + \frac{|a||x-y|}{1-|a|r}\right)\left(1 + \frac{|a|(1-r)}{1+|a|r}\right)\right)}{\log\left(1 + \frac{|x-y|}{1-r}\right)}$$

$$= \frac{\log\left(1 + \frac{2r\theta}{1-r}\left(1 + \frac{2|a|r\theta}{1-|a|r}\right)\left(1 + \frac{|a|(1-r)}{1+|a|r}\right)\right)}{\log\left(1 + \frac{2r\theta}{1-r}\right)},$$

where $\theta = \frac{|x-y|}{2r}$.

By Lemma 4.7 it follows that

$$J(x, y; a) \leq \frac{\log(1 + \frac{2r\theta}{1-r}(1 + \frac{|a|(1-r)}{1-|a|r} + \frac{2|a|r\theta}{1-|a|r}))}{\log\left(1 + \frac{2r\theta}{1-r}\right)}$$

$$= 1 + \frac{\log(1 + \frac{2|a|r\theta}{1-|a|r})}{\log\left(1 + \frac{2r\theta}{1-r}\right)} \le 1 + \frac{\log(1 + \frac{2|a|r}{1-|a|r})}{\log\left(1 + \frac{2r}{1-r}\right)}$$

$$= 1 + \frac{\text{arth}(|a|r)}{\text{arth}r} \le 1 + |a|.$$

Therefore, we get

$$j_{\mathbb{B}^n}(f(x), f(y)) \le (1 + |a|) j_{\mathbb{B}^n}(x, y).$$

That the upper bound $1 + |a|$ is sharp is already proved in [83].

Looking at the left-hand side of inequality, note that $f^{-1}(x) = A^{-1}\sigma_a^{-1}(x) = A^{-1}\sigma_a(x)$, where $\sigma_a(x)$ and A are as above. Note that since A is an orthogonal matrix, so is A^{-1}. Applying the above proof, for $x, y \in \mathbb{B}^n$, we get

$$\frac{j_{\mathbb{B}^n}(f^{-1}(x), f^{-1}(y))}{j_{\mathbb{B}^n}(x, y)} = \frac{j_{\mathbb{B}^n}(\sigma_a(x), \sigma_a(y))}{j_{\mathbb{B}^n}(x, y)} \le 1 + |a|.$$

Therefore, we have

$$j_{\mathbb{B}^n}(f(x), f(y)) \ge \frac{1}{1 + |a|} j_{\mathbb{B}^n}(x, y). \tag{4.9}$$

$\square$

Lemma 4.8 *For $a \ge 0, q \in [0, 1]$, we have*

$$\log\left(\frac{q + e^a}{1 + qe^a}\right) \le \frac{1 - q}{1 + q}a.$$

Proof Denote

$$f(a, q) := \log\left(\frac{q + e^a}{1 + qe^a}\right) - \frac{1 - q}{1 + q}a.$$

By differentiation, we have

$$f_a'(a, q) = -\frac{q(1 - q)}{1 + q} \frac{(e^a - 1)^2}{(1 + qe^a)(q + e^a)}.$$

Therefore we conclude that

$$f(a, q) \le f(0, q) = 0.$$

$\square$

Looking at the definition of the distance ratio metric, it is natural to expect that some properties of the logarithm will be essential for establishing properties of this metric. In fact, in [147], the classical Bernoulli inequality [158, (3.6)] was used for this purpose. However, it seems that now some stronger inequalities are needed and in particular we need to use the following result, which gives good estimates whenever the Lipschitz constant C satisfies $1 \leq C \leq 2$, and it also allows us to eliminate logarithms in further considerations.

Theorem 4.9 *Let D and D' be proper subdomains of $\mathbb{R}^n$. For an open continuous mapping $f : D \to D'$ denote*

$$X = X(z, w) := \frac{|z - w|}{\min\{d_D(z), d_D(w)\}};$$

$$Y = Y(z, w) := \frac{|z - w|}{|f(z) - f(w)|} \frac{\min\{d_{D'}(f(z)), d_{D'}(f(w))\}}{\min\{d_D(z), d_D(w)\}}.$$

If there exists a number q, $0 \leq q \leq 1$, such that

$$q \leq Y + \frac{Y - 1}{X + 1}, \tag{4.10}$$

then the inequality

$$j_{D'}(f(z), f(w)) \leq \frac{2}{1 + q} j_D(z, w),$$

holds for all $z, w \in D$.

Proof Observe that

$$X = \frac{|z - w|}{\min\{d_D(z), d_D(w)\}} = \exp(j_D(z, w)) - 1,$$

and

$$Y = \frac{|z - w|}{|f(z) - f(w)|} \frac{\min\{d_{D'}(f(z)), d_{D'}(f(w))\}}{\min\{d_D(z), d_D(w)\}} = \frac{\exp(j_D(z, w)) - 1}{\exp(j_{D'}(f(z), f(w))) - 1}.$$

Hence the condition (4.10) is equivalent to

$$\exp(j_{D'}(f(z), f(w))) \leq \exp(j_D(z, w)) \left(\frac{q + e^{j_D(z, w)}}{1 + q e^{j_D(z, w)}} \right).$$

Therefore, by Lemma 4.8, we get

$$j_{D'}(f(z), f(w)) \le j_D(z, w) + \log \left(\frac{q + e^{j_D(z,w)}}{1 + q e^{j_D(z,w)}} \right)$$

$$\le j_D(z, w) + \frac{1-q}{1+q} j_D(z, w) = \frac{2}{1+q} j_D(z, w).$$

□

Lemma 4.10 ([146, Lemma 2.4]) *For positive numbers A, B, D and $0 < C < 1, \theta \ge 0$, we have the following:*

1. *The inequality*

$$1 + \frac{B}{D}\theta \left(1 + \frac{D}{1+A} \right) \left(1 + \frac{B}{1-C}\theta \right) \le \left(1 + \frac{B}{D}\theta \right) \left(1 + \frac{B}{1-C}\theta \right),$$

   holds if and only if $B\theta \le A + C$;
2. *The function*

$$\frac{\log(1 + \frac{B}{1-C}\theta)}{\log(1 + \frac{B}{D}\theta)}$$

   is monotone increasing (decreasing) in θ if $C + D < 1$ $(C + D > 1)$.

Proof The proof of the first part uses a simple direct argument and is therefore left to the reader. To prove the second part, let

$$f_1(\theta) = \log \left(1 + \frac{B}{1-C}\theta \right), \ f_1(0) = 0; \quad f_2(\theta) = \log \left(1 + \frac{B}{D}\theta \right), \ f_2(0) = 0.$$

Since

$$\frac{f_1'(\theta)}{f_2'(\theta)} = \frac{D + B\theta}{1 - C + B\theta} = 1 + \frac{C + D - 1}{1 - C + B\theta},$$

the proof follows from Lemma 4.6. □

The following theorem was first conjectured in [147] in the special case of the punctured disk.

Theorem 4.11 ([146]) *Let $a \in \mathbb{B}^n$ and $h : \mathbb{B}^n \to \mathbb{B}^n = h(\mathbb{B}^n)$ be a Möbius transformation with $h(0) = a$. Then $h(\mathbb{B}^n \setminus \{0\}) = \mathbb{B}^n \setminus \{a\}$ and for $x, y \in \mathbb{B}^n \setminus \{0\}$*

$$j_{\mathbb{B}^n \setminus \{a\}}(h(x), h(y)) \le C(a) j_{\mathbb{B}^n \setminus \{0\}}(x, y),$$

where the constant $C(a) = 1 + (\log \frac{2+|a|}{2-|a|})/\log 3$ is best possible.

Note that $C(a) < 1 + |a| < 2$ for all $a \in \mathbb{B}^n$, and hence the constant $C(a)$ is smaller than the constant $1 + |f(0)|$ in Theorem 4.3 and far smaller than the constant 2 in Theorem 4.2.

Proof Without loosing generality, we may assume that $h(z) = \sigma_a(z)$ and suppose in the rest of the proof that $|z| \geq |w|$. Let $G = \mathbb{B}^n \setminus \{0\}$ and $G' = \mathbb{B}^n \setminus \{a\}$. Then

$$
\begin{aligned}
j_G(z, w) &= \log\left(1 + \frac{|z - w|}{\min\{|z|, |w|, 1 - |z|, 1 - |w|\}}\right) \\
&= \log\left(1 + \frac{|z - w|}{\min\{|w|, 1 - |z|\}}\right),
\end{aligned}
$$

and

$$
j_{G'}(\sigma_a(z), \sigma_a(w)) = \log\left(1 + \frac{|\sigma_a(z) - \sigma_a(w)|}{T}\right),
$$

where

$$
T = T_a(z, w) := \min\{|\sigma_a(z) - a|, |\sigma_a(w) - a|, 1 - |\sigma_a(z)|, 1 - |\sigma_a(w)|\}.
$$

Depending on the values of the number T, the proof is divided into four cases. We consider each case separately, applying the Bernoulli inequality in the first case, the assertion from Theorem 4.9 in the second case, and a direct approach in the last two cases.

1. $T = |\sigma_a(z) - a|$.

   Since $|\sigma_a(z) - a| = |\sigma_a(z) - \sigma_a(0)| = \frac{s^2|z|}{|a^*||z-a^*|}$ and $|\sigma_a(z) - \sigma_a(w)| = \frac{s^2|z-w|}{|z-a^*||w-a^*|}$, we have

   $$
   j_{G'}(\sigma_a(z), \sigma_a(w)) = \log\left(1 + \frac{|z - w|}{|a||z||w - a^*|}\right).
   $$

   Suppose first that $|w| \leq 1 - |z|$. Since also $|w| \leq 1 - |z| \leq 1 - |w|$, we get that $0 \leq |w| \leq 1/2$. Therefore, by the Bernoulli inequality (see, e.g., [158, (3.6)]), we get

   $$
   j_{G'}(\sigma_a(z), \sigma_a(w)) \leq \log\left(1 + \frac{|z - w|}{|z|(1 - |a||w|)}\right) \leq \log\left(1 + \frac{|z - w|}{|w|(1 - \frac{|a|}{2})}\right)
   $$

   $$
   \leq \frac{1}{1 - \frac{|a|}{2}} \log\left(1 + \frac{|z - w|}{|w|}\right) = \frac{1}{1 - \frac{|a|}{2}} j_G(z, w).
   $$

   Assume now $1 - |z| \leq |w|(\leq |z|)$. Then $1/2 \leq |z| < 1$.

Note that in this case $(|z| - \frac{1}{2})(2 - |a|(1 + |z|)) \geq 0$, so we obtain that

$$\frac{1}{|z|(1 - |a||z|)} \leq \frac{1}{(1 - \frac{|a|}{2})(1 - |z|)}.$$

Hence,

$$j_{G'}(\sigma_a(z), \sigma_a(w)) \leq \log\left(1 + \frac{|z - w|}{|z|(1 - |a||w|)}\right) \leq \log\left(1 + \frac{|z - w|}{|z|(1 - |a||z|)}\right)$$

$$\leq \log\left(1 + \frac{|z - w|}{(1 - \frac{|a|}{2})(1 - |z|)}\right) \leq \frac{1}{1 - \frac{|a|}{2}} \log\left(1 + \frac{|z - w|}{1 - |z|}\right) = \frac{1}{1 - \frac{|a|}{2}} j_G(z, w).$$

2. $T = |\sigma_a(w) - a|$.

This case can be considered using Theorem 4.9 with the same resulting constant $C_1(a) = \frac{2}{2 - |a|}$. Namely, in terms of Theorem 4.9, we consider first the case $|w| \leq 1 - |z|$.

We get

$$X = \frac{|z - w|}{|w|} \geq \frac{|z| - |w|}{|w|} = \frac{|z|}{|w|} - 1 = X^*,$$

and

$$Y = \frac{|z - a^*|}{|a^*|} \geq 1 - |a||z| = Y^*.$$

Therefore,

$$Y + \frac{Y - 1}{X + 1} \geq Y^* - \frac{1 - Y^*}{1 + X^*} = 1 - |a|(|w| + |z|) \geq 1 - |a| = q.$$

In the second case, i.e., when $1 - |z| \leq |w|$, we want to show that

$$Y + \frac{Y - 1}{X + 1} \geq 1 - |a|$$

which is equivalent to $(Y - (1 - |a|))(1 + X) + Y \geq 1$. Since in this case

$$X = \frac{|z - w|}{1 - |z|} \geq \frac{|z| - |w|}{1 - |z|} := X^*$$

and

$$Y = \frac{|w||z - a^*|}{|a^*|(1 - |z|)} \geq (1 - |a||z|)\frac{|w|}{1 - |z|}$$

$$= 1 - |a||z| + (|w| + |z| - 1)\frac{1 - |a||z|}{1 - |z|} := Y^*,$$

we get

$$(Y - (1 - |a|))(1 + X) + Y - 1 \geq (Y^* - (1 - |a|))(1 + X^*) + Y^* - 1$$

$$= \left[|a|(1-|z|)+(|w|+|z|-1)\frac{1-|a||z|}{1-|z|}\right]\frac{1-|w|}{1-|z|}-|a||z|+(|w|+|z|-1)\frac{1-|a||z|}{1-|z|}$$

$$\geq |a|(1 - |w| - |z|) + (|w| + |z| - 1)\frac{1 - |a||z|}{1-|z|} = (|w| + |z| - 1)\frac{1 - |a|}{1 - |z|} \geq 0.$$

Hence, by Theorem 4.9, in both cases we get

$$j_{G'}(\sigma_a(z), \sigma_a(w)) \leq \frac{2}{1+q}j_G(z, w) = \frac{2}{2 - |a|}j_G(z, w) = C_1(a)j_G(z, w).$$

3. $T = 1 - |\sigma_a(z)|$.

   In this case, combining a well-known fact which can be found, for example, in [147, Lemma 3.2], Lemma 4.4 above, the facts

$$|a|^2|z - a^*|^2 - |z - a|^2 = (1 - |a|^2)(1 - |z|^2) \text{ and } |\sigma_a(z)| \leq \frac{|a| + |z|}{1 + |a||z|},$$

and the inequalities $|a||w - a^*| \geq 1 - |a||w| \geq 1 - |a||z|$, we get

$$j_{G'}(\sigma_a(z), \sigma_a(w)) = \log\left(1 + \frac{|\sigma_a(z) - \sigma_a(w)|}{1 - |\sigma_a(z)|}\right)$$

$$= \log\left(1 + \frac{|a|s^2|z - w|}{|w - a^*|(|a||z - a^*| - |z - a|)}\right)$$

$$= \log\left(1 + \frac{|z - w|(|a||z - a^*| + |z - a|)}{|a||w - a^*|(1 - |z|^2)}\right)$$

$$= \log\left(1 + \frac{|z - w|}{1 - |z|^2}\left|\frac{z - a^*}{w - a^*}\right|\left(1 + \frac{|z - a|}{|a||z - a^*|}\right)\right)$$

$$\leq \log\left(1 + \frac{|z - w|}{1 - |z|^2}\left(1 + \frac{|z - w|}{|w - a^*|}\right)\left(1 + \frac{|a| + |z|}{1 + |a||z|}\right)\right)$$

$$\leq \log\left(1 + \frac{|z - w|}{1 - |z|}\left(1 + \frac{|a||z - w|}{1 - |a||z|}\right)\left(1 + \frac{|a|(1 - |z|)}{1 + |a||z|}\right)\right).$$

We use here Lemma 4.10 part 1, together with the equalities

$$A = |a||z|, \ B = |a|, \ C = |a||w|, \ D = |a|(1 - |z|), \ \theta = |z - w|,$$

we obtain

$$j_{G'}(\sigma_a(z), \sigma_a(w)) \le \log\left[\left(1 + \frac{|z-w|}{1-|z|}\right)\left(1 + \frac{|a||z-w|}{1-|a||w|}\right)\right]. \qquad (4.11)$$

Assume that $1 - |z| \le |w|$ ($\le |z|$). Using Lemma 4.10 part 2 and the equalities

$$B = |a|, C = |a||z|, D = |a|(1 - |z|), \theta = |z - w|,$$

we get

$$J(z, w; a) := \frac{j_{G'}(\sigma_a(z), \sigma_a(w))}{j_G(z, w)} \le 1 + \frac{\log(1 + \frac{|a||z-w|}{1-|a||z|})}{\log\left(1 + \frac{|z-w|}{1-|z|}\right)}$$

$$\le 1 + \frac{\log(1 + \frac{2|a||z|}{1-|a||z|})}{\log\left(1 + \frac{2|z|}{1-|z|}\right)},$$

since in this case, we have $C + D = |a| < 1$ and $|z - w| \le 2|z|$. Since the last function is monotone decreasing in $|z|$ and $|z| \ge 1/2$, we get

$$J(z, w; a) \le 1 + \frac{\log(\frac{1+\frac{1}{2}|a|}{1-\frac{1}{2}|a|})}{\log\left(\frac{1+\frac{1}{2}}{1-\frac{1}{2}}\right)} = 1 + (\log\frac{2+|a|}{2-|a|})/\log 3 := C_2(a).$$

Now suppose $|w| \le 1 - |z|(\le 1 - |w|)$. The estimate (4.11) and Lemma 4.10 part 2, with

$$B = |a|, \ C = D = |a||w|, \ \theta = |z - w|,$$

yield

$$J(z, w; a) \le \frac{\log\left[\left(1 + \frac{|z-w|}{1-|z|}\right)\left(1 + \frac{|a||z-w|}{1-|a||w|}\right)\right]}{\log\left(1 + \frac{|z-w|}{|w|}\right)}$$

$$\le \frac{\log\left[\left(1 + \frac{|z-w|}{|w|}\right)\left(1 + \frac{|a||z-w|}{1-|a||w|}\right)\right]}{\log\left(1 + \frac{|z-w|}{|w|}\right)}$$

$$= 1 + \frac{\log\left(1 + \frac{|a||z-w|}{1-|a||w|}\right)}{\log\left(1 + \frac{|z-w|}{|w|}\right)} \le 1 + \frac{\log\left(1 + \frac{|a|}{1-|a||w|}\right)}{\log\left(1 + \frac{1}{|w|}\right)},$$

since $C + D = 2|a||w| \leq |a| < 1$ and $0 \leq |z - w| \leq |z| + |w| \leq 1$.

Let us denote the last function as $g(|w|)$ and let $|w| = r$, $0 < r \leq 1/2$. Since

$$g'(r) = \frac{|a|^2}{(1 - r|a|)(1 + (1 - r)|a|)\log(1 + 1/r)}$$

$$+ \frac{\log\left(1 + \frac{|a|}{1-|a|r}\right)}{r(1 + r)\log^2(1 + 1/r)} > 0,$$

it follows that $g(r)$ is a monotone increasing function and we finally obtain

$$J(z, w; a) \leq 1 + \frac{\log\left(1 + \frac{|a|}{1-|a|/2}\right)}{\log(1 + 2)} = C_2(a).$$

4. $T = 1 - |\sigma_a(w)|$.

This case is treated similar to the previous case,

$$j_{G'}(\sigma_a(z), \sigma_a(w)) = \log\left(1 + \frac{|a|s^2|z - w|}{|z - a^*|(|a||w - a^*| - |w - a|)}\right)$$

$$= \log\left(1 + \frac{|z - w|(|a||w - a^*| + |w - a|)}{|a||z - a^*|(1 - |w|^2)}\right)$$

$$\leq \log\left(1 + \frac{|z - w|}{1 - |w|}\left(1 + \frac{|a||z - w|}{1 - |a||z|}\right)\left(1 + \frac{|a|(1 - |w|)}{1 + |a||w|}\right)\right).$$

Applying Lemma 4.10 part 1 with

$$A = |a||w|, \ B = |a|, \ C = |a||z|, \ D = |a|(1 - |w|), \ \theta = |z - w|,$$

we obtain

$$j_{G'}(\sigma_a(z), \sigma_a(w)) \leq \log\left[\left(1 + \frac{|z - w|}{1 - |w|}\right)\left(1 + \frac{|a||z - w|}{1 - |a||z|}\right)\right]. \tag{4.12}$$

Assume that $1 - |z| \leq |w|(\leq |z|)$. We get

$$J(z, w; a) := \frac{j_{G'}(\sigma_a(z), \sigma_a(w))}{j_G(z, w)} \leq \frac{\log\left[\left(1 + \frac{|z-w|}{1-|z|}\right)\left(1 + \frac{|a||z-w|}{1-|a||z|}\right)\right]}{\log\left(1 + \frac{|z-w|}{1-|z|}\right)}$$

$$= 1 + \frac{\log(1 + \frac{|a||z-w|}{1-|a||z|})}{\log\left(1 + \frac{|z-w|}{1-|z|}\right)},$$

and we have already considered this inequality above. In the case $|w| \leq 1 - |z| \leq 1 - |w|$, we have

$$J(z, w; a) \leq \frac{\log\left[\left(1 + \frac{|z-w|}{1-|w|}\right)\left(1 + \frac{|a||z-w|}{1-|a||z|}\right)\right]}{\log\left(1 + \frac{|z-w|}{|w|}\right)}$$

$$\leq \frac{\log\left[\left(1 + \frac{|z-w|}{|w|}\right)\left(1 + \frac{|a||z-w|}{1-|a|(1-|w|)}\right)\right]}{\log\left(1 + \frac{|z-w|}{|w|}\right)}$$

$$= 1 + \frac{\log\left(1 + \frac{|a||z-w|}{1-|a|(1-|w|)}\right)}{\log\left(1 + \frac{|z-w|}{|w|}\right)} \leq 1 + \frac{\log\left(1 + \frac{|a|}{1-|a|(1-|w|)}\right)}{\log\left(1 + \frac{1}{|w|}\right)},$$

where the last inequality follows from Lemma 4.10 part 2 for

$$B = |a|, \quad C = |a|(1 - |w|), \quad D = |a||w|, \quad \theta = |z - w|,$$

since $C + D = |a| < 1$ and $|z - w| \leq |z| + |w| \leq 1$.
  Put now $|w| = r$ and let $k(r) = k_1(r)/k_2(r)$ where

$$k_1(r) = \log\left(1 + \frac{|a|}{1 - |a|(1 - r)}\right); \quad k_2(r) = \log\left(1 + \frac{1}{r}\right).$$

Let us show now that the function $k(r)$ is monotone increasing on the positive part of the real axis. Namely, since $k_1(\infty) = k_2(\infty) = 0$ and

$$k_1'(r)/k_2'(r) = \frac{|a|^2 r(1 + r)}{(1 + |a|r)(1 - |a| + |a|r)} = \frac{|a|(1 + r)}{1 + |a|r} \frac{|a|r}{1 - |a| + |a|r}$$

$$= \left(1 - \frac{1 - |a|}{1 + |a|r}\right)\left(1 - \frac{1 - |a|}{1 - |a| + |a|r}\right),$$

with both functions in parenthesis increasing on $\mathbb{R}^+$, the conclusion follows from Lemma 4.6. This is so because in this case $0 < r \leq 1/2$, we also obtain that

$$J(z, w; a) \leq 1 + \frac{\log\left(1 + \frac{|a|}{1-|a|/2}\right)}{\log\left(1 + 2\right)} = C_2(a).$$

In order to verify that the constant $C_2(a)$ is sharp, we first calculate $T_a(\frac{a}{2|a|}, \frac{-a}{2|a|}) = \frac{1-|a|}{2+|a|}$. Using this, it follows that

$$J\left(\frac{a}{2|a|}, \frac{-a}{2|a|}; a\right) = C_2(a).$$

Since $C_2(a) = 1 + (\log\frac{2+|a|}{2-|a|})/\log 3 > \frac{1}{1-\frac{|a|}{2}} = C_1(a)$, we conclude that the best possible upper bound C is $C = C_2(a)$.

□

Since proper domains for Möbius transformation are $\mathbb{B}^n$ and $\mathbb{H}^n$, it follows that for dimension $n = 2$, much more can be said. For example, as a special case $K = 1$ of Theorem 2.45, we obtain Proposition 4.12 below. This is so because in a half plane $\mathbb{H}^2$, by [24, Theorem 7.2.1, p. 130], we have

$$\mathrm{th}\left(\frac{\rho_H(z, w)}{2}\right) = \frac{|z - w|}{|z - \overline{w}|}.$$

As another example, in dimension $n = 2$, one can prove an analogue of Theorem 4.2, valid for *arbitrary holomorphic mappings* of a half-plane into itself. Hence in this case we get a substantial generalization of Theorem 4.2. The main tool in the proof of such generalization will be the following Julia's variant of the famous Schwarz–Pick Lemma for the half-plane [26].

Proposition 4.12 *For all holomorphic mappings* f, $f : \mathbb{H}^2 \to \mathbb{H}^2$, *and all* $z, w \in \mathbb{H}^2$, *we have*

$$\left|\frac{f(z) - f(w)}{f(z) - \overline{f(w)}}\right| \le \left|\frac{z - w}{z - \overline{w}}\right|.$$

Proof As indicated above, this follows from Theorem 2.45 when $K = 1$.   □

Because the j-metric is invariant under similarities, we can restrict the proof to the case of the upper half-plane $\mathbb{H}^2 := \{z : \mathrm{Im}\, z > 0\}$. A generalization of Theorem 4.2 for this domain is given by the following theorem.

Theorem 4.13 ([144]) *If* $f : \mathbb{H}^2 \to \mathbb{H}^2$ *is a holomorphic mappings and* $z, w \in \mathbb{H}^2$, *then*

$$j_{\mathbb{H}^2}(f(z), f(w)) \le 2 j_{\mathbb{H}^2}(z, w),$$

where the constant 2 is best possible.

Proof Denote $s = \min\{\mathrm{Im}\, z, \mathrm{Im}\, w\}$ and suppose that $\mathrm{Im}\, f(z) \le \mathrm{Im}\, f(w)$. Then

$$j_{\mathbb{H}^2}(z, w) = \log\left(1 + \frac{|z - w|}{s}\right); \quad j_{\mathbb{H}^2}(f(z), f(w)) = \log\left(1 + \frac{|f(z) - f(w)|}{\operatorname{Im} f(z)}\right).$$

Applying Proposition 4.12, we get

$$\left|\frac{f(z) - \overline{f(w)}}{f(z) - f(w)}\right|^2 - 1 \geq \left|\frac{z - \overline{w}}{z - w}\right|^2 - 1,$$

that is,

$$\left|\frac{f(z) - f(w)}{z - w}\right|^2 \leq \frac{\operatorname{Im} f(z)\operatorname{Im} f(w)}{\operatorname{Im} z\operatorname{Im} w},$$

since for all $x, y \in \mathbb{C}$, the identity $|x - \overline{y}|^2 - |x - y|^2 = 4\operatorname{Im} x\operatorname{Im} y$ holds. Therefore,

$$\frac{|f(z) - f(w)|^2}{\operatorname{Im}^2 f(z)} \leq \frac{|z - w|^2}{s^2}\frac{\operatorname{Im} f(w)}{\operatorname{Im} f(z)} = \frac{|z - w|^2}{s^2}(1 + \frac{\operatorname{Im}(f(w) - f(z))}{\operatorname{Im} f(z)}),$$

and so

$$\frac{|f(z) - f(w)|}{\operatorname{Im} f(z)} \leq \frac{|z - w|}{s}\sqrt{1 + \frac{|f(z) - f(w)|}{\operatorname{Im} f(z)}}, \qquad (*)$$

since $\operatorname{Im} x \leq |x|, x \in \mathbb{C}$.

Denote $\frac{|z-w|}{s} := 2X$ and $1 + \frac{|f(z)-f(w)|}{\operatorname{Im} f(z)} := Y$. From the relation $(*)$ we get that $Y - 2X\sqrt{Y} \leq 1$, and so $(\sqrt{Y} - X)^2 \leq 1 + X^2$ and $\sqrt{Y} \leq X + \sqrt{1 + X^2}$, which is the same as

$$1 + \frac{|f(z) - f(w)|}{\operatorname{Im} f(z)} \leq (X + \sqrt{1 + X^2})^2.$$

Hence

$$j_{\mathbb{H}^2}(f(z), f(w)) = \log\left(1 + \frac{|f(z) - f(w)|}{\operatorname{Im} f(z)}\right)$$
$$\leq 2\log(X + \sqrt{1 + X^2}) \leq 2\log(1 + 2X) = 2j_{\mathbb{H}^2}(z, w),$$

and the proof is complete.

In order to establish that the constant 2 is best possible, choose

$$f_0(z) = a - \frac{1}{b + z},$$

where a and b are arbitrary real numbers. Since $\operatorname{Im} f_0(z) = \frac{\operatorname{Im} z}{|b+z|^2}$, it follows that $f_0 : \mathbb{H}^2 \to \mathbb{H}^2$. A calculation of j values along the line $\zeta \subset \mathbb{H}^2$ given by

$$\zeta := \{z = i - b + t, t \in \mathbb{R}\}$$

with $w = i - b$ yields

$$j_{\mathbb{H}^2}(z, w) = \log(1+t); \quad j_{\mathbb{H}^2}(f_0(z), f_0(w)) = \log(1+\frac{\frac{t}{\sqrt{1+t^2}}}{\frac{1}{1+t^2}}) = \log(1+t\sqrt{1+t^2}).$$

Hence,

$$\frac{j_{\mathbb{H}^2}(f_0(z), f_0(w))}{j_{\mathbb{H}^2}(z, w)} = \frac{\log(1 + t\sqrt{1+t^2})}{\log(1+t)},$$

and this expression tends to 2 as $t \to \infty$.

Therefore the constant 2 is best possible. $\qquad\square$

A similar result for the unit disk $\mathbb{D}$ is given in [145].

Theorem 4.14 *For every holomorphic mapping $f : \mathbb{D} \to \mathbb{D}$ and all $z, w \in \mathbb{D}$, we have*

$$j_{\mathbb{D}}(f(z), f(w)) \leq 2 j_{\mathbb{D}}(z, w).$$

Remark 4.15 It cannot be claimed in this case that the best possible Lipschitz constant is $C^* = 2$, since an example of an analytic function $f_0 : \mathbb{D} \to \mathbb{D}$ which satisfies

$$\sup_{z,w \in \mathbb{D}} \frac{j_{\mathbb{D}}(f_0(z), f_0(w))}{j_{\mathbb{D}}(z, w)} = 2$$

is not known.

This question was treated in [142] where the following estimation for $C^* = C^*(a)$ is proved:

$$1 + |a| \leq C^*(a) \leq \min\{2(1 + |a|), \sqrt{5 + 2|a| + |a|^2}\}; \quad a := f(0).$$

By Theorem 4.14, this reduces to

$$1 + |a| \leq C^*(a) \leq 2,$$

but the problem of determining the best possible Lipschitz constant in this case remains open.

4.2 Lipschitz Continuity and Analytic Functions

We begin this section by defining the sector as follows:

$$S_\varphi = \{re^{i\theta} \in \mathbb{C} : 0 < \theta < \varphi, r > 0\}.$$

Lemma 4.16 ([147]) *Let $k \in (1, \infty)$, $\theta \in (0, \frac{\pi}{2k})$ and $r \in (0, 1)$. Then*

1. $f_1(\theta) \equiv \frac{k \sin \theta}{1 - \sin \theta} - \frac{\sin k\theta}{1 - \sin k\theta}$ *is decreasing from $(0, \frac{\pi}{2k})$ to $(-\infty, 0)$.*

2. $f_2(r) \equiv \frac{1 - (1 - \sin \theta)r}{1 - (1 - \sin k\theta)r^k}$ *is decreasing from $(0, 1)$ to $(\frac{\sin \theta}{\sin k\theta}, 1)$.*

3. $f_3(r) \equiv \frac{\log(1 + \frac{1 - r^k}{r^k \sin k\theta})}{\log(1 + \frac{1 - r}{r \sin \theta})}$ *is decreasing from $(0, 1)$ to $(\frac{k \sin \theta}{\sin k\theta}, k)$.*

Proof

1. Differentiating, we get

$$f_1'(\theta) = k[g(1) - g(k)],$$

where $g(k) = \frac{\cos k\theta}{(1 - \sin k\theta)^2}$. Since

$$g'(k) = \frac{\theta(1 - \sin k\theta + \cos^2 k\theta)}{(1 - \sin k\theta)^3} > 0,$$

we have $f_1'(\theta) < 0$ and hence f_1 is decreasing in $(0, \frac{\pi}{2k})$. The limiting values are clear.

2. Differentiating again, we get

$$[1 - (1 - \sin k\theta)r^k]^2 f_2'(r)$$
$$= -(k - 1)(1 - \sin \theta)(1 - \sin k\theta)r^k + k(1 - \sin k\theta)r^{k-1} - (1 - \sin \theta)$$
$$\equiv h(r).$$

Since

$$h'(r) = k(k - 1)(1 - \sin k\theta)r^{k-2}[1 - r(1 - \sin \theta)] > 0,$$

we have that h is increasing in $(0, 1)$ and hence $h(r) \leq h(1)$. By (1),

$$h(1) = (1 - \sin \theta)(1 - \sin k\theta)f_1(\theta) < 0.$$

Therefore $f_2'(r) < 0$, and hence f_2 is decreasing in $(0, 1)$. The limiting values are also clear.

3. Let $f_3(r) \equiv \frac{g_1(r)}{g_2(r)}$, where $g_1(r) = \log(1 + \frac{1-r^k}{r^k \sin k\theta})$ and $g_2(r) = \log(1 + \frac{1-r}{r \sin \theta})$. Then $g_1(1^-) = g_2(1^-) = 0$. Differentiating, we have

$$\frac{g_1'(r)}{g_2'(r)} = kf_2(r).$$

Therefore, f_3 is increasing in $(0, 1)$ by Lemma 4.6. By l'Hôpital's rule and (2) we get the limiting values easily.

$\square$

Lemma 4.17 ([147]) *Let $f : S_{\pi/k} \to \mathbb{H}^2$ with $f(z) = z^k$ ($k \geq 1$). Let $x, y \in S_{\pi/k}$ with $\arg(x) = \arg(y) = \theta$. Then*

$$\frac{k \sin \theta}{\sin k\theta} j_{S_{\pi/k}}(x, y) \leq j_{\mathbb{H}^2}(f(x), f(y)) \leq kj_{S_{\pi/k}}(x, y).$$

Proof Since the j-metric is invariant under similarity, it can be assumed that $x = re^{i\theta}$ and $y = e^{i\theta}$, $0 < r < 1$. By symmetry, we also assume $0 < \theta \leq \frac{\pi}{2k}$. Then

$$\frac{j_{\mathbb{H}^2}(f(x), f(y))}{j_{S_{\pi/k}}(x, y)} = \frac{\log(1 + \frac{1-r^k}{r^k \sin k\theta})}{\log(1 + \frac{1-r}{r \sin \theta})}.$$

By Lemma 4.16(3), the result follows.

$\square$

Lemma 4.18 ([147]) *Let $n \in \mathbb{N}$, $0 < \theta \leq \frac{\pi}{2n}$. Then for $x, y \in \mathbb{R}^n \setminus \{0\}$,*

$$1 + \frac{|x^n - y^n|}{|x|^n \sin n\theta} \leq (1 + \frac{|x - y|}{|x| \sin \theta})^n.$$

Proof It is clear that (4.18) holds if $n = 1$. Next, suppose that (4.18) holds when $n = k$. Namely,

$$1 + \frac{|x^k - y^k|}{|x|^k \sin k\theta} \leq (1 + \frac{|x - y|}{|x| \sin \theta})^k, \tag{4.13}$$

where $0 < \theta \leq \frac{\pi}{2k}$. Then, if $n = k+1$, we have $0 < \theta \leq \frac{\pi}{2(k+1)} < \frac{\pi}{2k}$ and by (4.13),

$$(1 + \frac{|x - y|}{|x| \sin \theta})^{k+1} \geq (1 + \frac{|x^k - y^k|}{|x|^k \sin k\theta})(1 + \frac{|x - y|}{|x| \sin \theta})$$

$$\geq 1 + \frac{|x - y|}{|x| \sin \theta} + \frac{|x^k - y^k|}{|x|^k \sin k\theta}(1 + \frac{|x - y|}{|x|})$$

$$\geq 1 + \frac{|x^{k+1} - x^k y|}{|x|^{k+1} \sin \theta} + \frac{|x^k y - y^{k+1}|}{|x|^{k+1} \sin k\theta}$$

$$\geq 1 + \frac{|x^{k+1} - y^{k+1}|}{|x|^{k+1} \sin(k+1)\theta}.$$

The proof is completed using the induction. □

Theorem 4.19 ([147]) *Let* $f : S_{\pi/k} \to \mathbb{H}^2$ *with* $f(z) = z^k$ ($k \in \mathbb{N}$). *Then for all* $x, y \in S_{\pi/k}$,

$$j_{\mathbb{H}^2}(f(x), f(y)) \leq kj_{S_{\pi/k}}(x, y),$$

and the constant k is the best possible.

Proof Using symmetry, we can assume that $d(f(x), \partial\mathbb{H}^2) \leq d(f(y), \partial\mathbb{H}^2)$. Using this and Lemma 4.18, we obtain

$$j_{\mathbb{H}^2}(f(x), f(y)) = \log(1 + \frac{|x^k - y^k|}{|x|^k \sin k\theta})$$

$$\leq k \log(1 + \frac{|x - y|}{|x| \sin \theta})$$

$$\leq kj_{S_{\pi/k}}(x, y),$$

where $0 < \theta = \min\{\arg(x), \frac{\pi}{k} - \arg(x)\} \leq \frac{\pi}{2k}$.

Let $x = re^{i\alpha}$ and $y = e^{i\alpha}$, where $0 < \alpha < \frac{\pi}{2k}$ and $0 < r < 1$. Letting $r \to 0$, by Lemma 4.16(3) and Lemma 4.17, we see that the constant k is the best possible. □

The sector problem concerning the j-metric has been also treated in [100] where the following specific question was stated.

Question 4.20 For $a, b \in (0, \pi)$ and $K \geq 1$, is there a constant C such that $C \to 1$ when $a \to b$ and $K \to 1$, and such that for every K-quasiconformal mapping $f : S_a \to S_b$ we have

$$j_{S_b}(f(a), f(b)) \leq C \cdot j_{S_a}(a, b)?$$

Although the constant C can be determined in some particular cases, the answer to the above question in general is not positive, as shown by the following example.

Let $S = S(\pi/2)$ and let φ be the inversion of S with respect to the unit circle $C = \{z| \, |z| = 1\}$. Let $z_1 = (\frac{\sqrt{2}}{2}, \frac{\sqrt{2}}{2})$, $z_2 = (\sqrt{3}, 1)$, $\omega_1 = \varphi(z_1) = z_1$, and $\omega_2 = \varphi(z_2) = (\frac{\sqrt{3}}{4}, \frac{1}{4})$.

Then a simple calculation shows that

$$j(z_1, z_2) \neq j(\omega_1, \omega_2).$$

Note that φ is harmonic and anticonformal, so $R \circ \varphi : S\left(\frac{\pi}{2}\right) \to S\left(\frac{\pi}{2}\right)$ is a conformal map, where R is reflection with respect to the line $x = y$.

Of course, $j(z_1, z_2) \neq j(R \circ \varphi(z_1), R \circ \varphi(z_2))$.

Lemma 4.21 ([143]) *For $z \in \mathbb{C}$ and $p \in \mathbb{N}$ we have*

$$\log(1 + |z^p - 1|) \le p \log(1 + |z - 1|).$$

Proof Putting $z = u + 1$, we obtain

$$1 + |z^p - 1| = 1 + \left|\sum_{k=1}^{p} \binom{p}{k} u^k\right| \le 1 + \sum_{k=1}^{p} \binom{p}{k} |u|^k = (1 + |u|)^p = (1 + |z - 1|)^p,$$

and the conclusion follows. □

Lemma 4.22 ([143]) *Let Q_d^* denote the class of all polynomials with exact degree d ($d \ge 1$) which have no zeros inside the disk $\mathbb{D}$.*

If $Q \in Q_d^*$ and $u, v \in \mathbb{D}$, then

$$\left|\frac{Q(u)}{Q(v)} - 1\right| \le \left(1 + \frac{|u - v|}{1 - |v|}\right)^d - 1.$$

Proof A polynomial $Q(z)$ with zeros $\{-b_k\}(k = 1, 2, \cdots, d)$ has a representation of the form

$$Q(z) = C \prod_{k=1}^{d} (z + b_k), C \neq 0.$$

Since $Q(z) \neq 0$ for $z \in \mathbb{D}$, necessarily $|b_k| \ge 1$. Therefore,

$$\left|\frac{Q(u)}{Q(v)} - 1\right| = \left|\prod_{k=1}^{d} \frac{u + b_k}{v + b_k} - 1\right| = \left|\prod_{k=1}^{d}\left(1 + \frac{u - v}{b_k + v}\right) - 1\right|$$

$$\le \prod_{k=1}^{d}\left(1 + \frac{|u - v|}{|b_k + v|}\right) - 1$$

$$\le \prod_{k=1}^{d}\left(1 + \frac{|u - v|}{|b_k| - |v|}\right) - 1$$

$$\le \left(1 + \frac{|u - v|}{1 - |v|}\right)^d - 1.$$

□

Remark 4.23 Note that since $|\frac{Q(u)}{Q(v)} - 1| \geq |\frac{Q(u)}{Q(v)}| - 1$, we also obtain

$$|\frac{Q(u)}{Q(v)}| \leq \left(1 + \frac{|u-v|}{1-|v|}\right)^d.$$

A result of [143] connecting the distance ratio metric with polynomial mappings can now be given.

Theorem 4.24 ([143]) *Let $p \in \mathbb{N}$ and $\{a_k\}$ be a sequence of complex numbers with $\sum_{k=1}^{p} |a_k| \leq 1$. Let $f : \mathbb{D} \setminus \{0\} \to \mathbb{D} \setminus \{0\}$ with $f(z) = \sum_{k=1}^{p} a_k z^k$ and $f(0) = 0$. Then for all $x, y \in \mathbb{D} \setminus \{0\}$*

$$j_{\mathbb{D}\setminus\{0\}}(f(x), f(y)) \leq p j_{\mathbb{D}\setminus\{0\}}(x, y),$$

and the constant p is sharp.

Proof For $x, y \in \mathbb{D} \setminus \{0\}$, we have

$$j_{\mathbb{D}\setminus\{0\}}(x, y) = \log(1 + \frac{|x-y|}{\min\{|x|, |y|, 1-|x|, 1-|y|\}})$$

and

$$j_{\mathbb{D}\setminus\{0\}}(f(x), f(y)) = \log(1 + \frac{|f(x) - f(y)|}{T}),$$

where $T = \min\{|f(x)|, |f(y)|, 1 - |f(x))|, 1 - |f(y)|\}$.

Case 1 $T = 1 - |f(x)|$. Noting that

$$|f(x) - f(y)| = |x - y||\sum_{k=1}^{p} a_k(\sum_{i+j=k-1} x^i y^j)| \leq |x - y| \sum_{k=1}^{\infty} |a_k|(\sum_{i=0}^{k-1} |x|^i)$$

and

$$1 - |f(x)| \geq \sum_{k=1}^{p} |a_k| - \sum_{k=1}^{p} |a_k||x|^k = (1 - |x|) \sum_{k=1}^{p} |a_k|(\sum_{i=0}^{k-1} |x|^i),$$

we obtain

$$j_{\mathbb{D}\setminus\{0\}}(f(x), f(y)) \leq \log(1 + \frac{|x-y|}{1-|x|}) \leq j_{\mathbb{D}\setminus\{0\}}(x, y).$$

Case 2 $T = 1 - |f(y)|$. This is similar to Case 1.

Case 3 $T = |f(y)|$. If 0 is an m-th order zero of f, since f obviously has no other zeros in $\mathbb{D}$, we get $f(z) = z^m Q(z)$, $Q \in Q_d^*$, $m + d = p$ (where Q_d^* is as in Lemma 4.22). By Lemma 4.22 and Remark 4.23, it follows that

$$\frac{|f(x) - f(y)|}{|f(y)|} = |\frac{x^m Q(x)}{y^m Q(y)} - 1| = |\left(\frac{x^m}{y^m} - 1\right)\frac{Q(x)}{Q(y)} + \frac{Q(x)}{Q(y)} - 1|$$

$$\leq |\frac{x^m}{y^m} - 1||\frac{Q(x)}{Q(y)}| + |\frac{Q(x)}{Q(y)} - 1|$$

$$\leq \left(1 + |\frac{x^m}{y^m} - 1|\right)\left(1 + \frac{|x - y|}{1 - |y|}\right)^d - 1.$$

Hence, by Lemma 4.21, we have

$$j_{\mathbb{D}\setminus\{0\}}(f(x), f(y)) = \log\left(1 + \frac{|f(x) - f(y)|}{|f(y)|}\right)$$

$$\leq d \log\left(1 + \frac{|x - y|}{1 - |y|}\right) + \log\left(1 + |\frac{x^m}{y^m} - 1|\right)$$

$$\leq d \log\left(1 + \frac{|x - y|}{1 - |y|}\right) + m \log(1 + \frac{|x - y|}{|y|})$$

$$\leq p j_{\mathbb{D}\setminus\{0\}}(x, y),$$

and the proof for Case 3 is complete

Case 4 $T = |f(x)|$. This case is similar to Case 3.

To establish the sharpness of the inequality, let $f(z) = z^p$ ($p \in \mathbb{N}$). For s, $t \in (0, \frac{1}{2})$ and $s < t$, we have

$$j_{\mathbb{D}\setminus\{0\}}(f(t), f(s)) = \log(1 + \frac{t^p - s^p}{s^p}) = p \log(\frac{t}{s}) = p j_{\mathbb{D}\setminus\{0\}}(t, s).$$

Therefore the constant p is sharp.

$\square$

It is of interest to investigate the Lipschitz continuity of the distance-ratio metric under some other conformal mappings. Let us present some results concerning analytic mappings of the unit disk $\mathbb{D}$ into itself. For example, for an analytic $f : \mathbb{D} \to \mathbb{D}$, supposing boundedness of l_1 norm of its Maclaurin coefficients, we give a proof of Lipschitz continuity with the best possible Lipschitz constant $C = 1$.

The following result from [142] gives a sufficient condition for an analytic mapping to be a contraction, i.e., to have the Lipschitz constant at most 1.

Theorem 4.25 *Let* $f : \mathbb{D} \to \mathbb{D}$ *be a nonconstant mapping given by* $f(z) = \sum_{k=0}^{\infty} a_k z^k$, *with the condition*

$$\sum_{k=0}^{\infty} |a_k| \leq 1. \qquad\qquad (*)$$

Then for all $x, y \in \mathbb{D}$,

$$j_{\mathbb{D}}(f(x), f(y)) \leq j_{\mathbb{D}}(x, y),$$

and this inequality is sharp.

Proof Suppose that $|f(x)| \geq |f(y)|$. Then

$$j_{\mathbb{D}}(x, y) = \log(1 + \frac{|x - y|}{\min\{1 - |x|, 1 - |y|\}})$$

and

$$j_{\mathbb{D}}(f(x), f(y)) = \log(1 + \frac{|f(x) - f(y)|}{1 - |f(x)|}).$$

We have that

$$|f(x) - f(y)| = |x - y| |\sum_{k=1}^{\infty} a_k(\sum_{i+j=k-1} x^i y^j)| \leq |x - y| \sum_{k=1}^{\infty} |a_k|(\sum_{i=0}^{k-1} |x|^i)$$

and

$$1 - |f(x)| \geq \sum_{k=1}^{\infty} |a_k| - \sum_{k=1}^{\infty} |a_k||x|^k = (1 - |x|) \sum_{k=1}^{\infty} |a_k|(\sum_{i=0}^{k-1} |x|^i).$$

Hence,

$$j_{\mathbb{D}}(f(x), f(y)) \leq \log(1 + \frac{|x - y|}{1 - |x|}) \leq j_{\mathbb{D}}(x, y).$$

Again in order to establish the sharpness of the inequality, let $a_p = 1, a_i = 0, i \neq p$, i.e., $f(z) = z^p$ ($p \in \mathbb{N}$). For $s, t \in (0, 1)$ and $s < t$, we have

$$\frac{j_{\mathbb{D}}(f(t), f(s))}{j_{\mathbb{D}}(t, s)} = \frac{\log \frac{1-s^p}{1-t^p}}{\log \frac{1-s}{1-t}} = \frac{\log \frac{1-s}{1-t} + \log \frac{1+s+\cdots+s^{p-1}}{1+t+\cdots+t^{p-1}}}{\log \frac{1-s}{1-t}}.$$

Letting $t \to 1^-$ we obtain $C = 1$. Therefore this constant is sharp. $\qquad \square$

Note that the condition $(*)$ is sufficient for f to map $\mathbb{D}$ into itself but is not necessary at all. For example, consider $f(z) = z + \frac{(1-z)^3}{4}$.

Indeed, applying the maximum modulus principle, we get

$$|f(z)| \leq \max_{\theta} |f(e^{i\theta})| = |e^{i\theta} + \frac{(1 - e^{i\theta})^3}{4}| = |e^{-i\theta/2} + \frac{(e^{-i\theta/2} - e^{i\theta/2})^3}{4}|$$

$$= |\cos \frac{\theta}{2} - i \sin \frac{\theta}{2}(1 - 2\sin^2 \frac{\theta}{2})| = |\cos \frac{\theta}{2} - i \sin \frac{\theta}{2} \cos \theta|$$

$$= \sqrt{\cos^2 \frac{\theta}{2} + \sin^2 \frac{\theta}{2} \cos^2 \theta} \leq \sqrt{\cos^2 \frac{\theta}{2} + \sin^2 \frac{\theta}{2}} = 1.$$

Chapter 5
Bi-Lipschitz Property of HQC Mappings

The inverse of a K-quasiconformal homeomorphism is also K-quasiconformal. By the Schwarz lemma for K-quasiconformal mappings we know that both mappings are Hölder continuous in the Euclidean metric with exponent $K^{1/(1-n)}$, and the Gehring–Osgood result yields the same conclusion in the quasihyperbolic metric. The class of harmonic K-quasiconformal interpolates between the classes of conformal maps and general quasiconformal maps. In this chapter we study the modulus of continuity of harmonic quasiconformal mappings relative to the quasihyperbolic metric and prove that both the mapping and its inverse are Lipschitz-continuous. Our presentation here is largely based on the papers [21, 99] and [103].

5.1 Bi-Lipschitz Property of HQC Mappings in Plane

If f is harmonic, then f^{-1} is not in general harmonic. In the influential paper [123] Pavlović showed that harmonic quasiconformal mappings of the unit disk D onto itself are bi-Lipschitz mappings. This paper has initiated an extensive line of research between the Lipschitz conditions and harmonic quasiconformal mappings (see, for example, [15, 73, 76, 79, 114], and the references therein).

The quasihyperbolic metric is invariant under Euclidean similarities, but it is not invariant under conformal mappings (not even under Möbius transformations). By a result of Gehring and Osgood [53], for each domain $\Omega \subseteq \mathbb{R}^n$ and points $x, y \in \Omega$ there exists a quasihyperbolic geodesic such that the quasihyperbolic metric is quasiinvariant under quasiconformal mappings (see Theorem 3.18). This means on large scales a quasiconformal mapping is Lipschitz with respect to the quasihyperblic metric and on small scales it is Hölder.

Recall that, for quasiconformal mappings, we have the following geometric notion of an average derivative (see [19, Definition 1.5]):

© Springer Nature Switzerland AG 2019
V. Todorčević, *Harmonic Quasiconformal Mappings and Hyperbolic Type Metrics*,
https://doi.org/10.1007/978-3-030-22591-9_5

$$\alpha_f(z) = \exp\left(\frac{1}{n}(\log J_f)_{B_z}\right). \tag{5.1}$$

Here

$$(\log J_f)_{B_z} = \frac{1}{m(B_z)}\int_{B_z}\log J_f \, dm, \quad B_z = B(z, d(z, \partial\Omega)).$$

In the case $n = 2$ we have

$$\frac{1}{\alpha_f(z)} = \exp\left(\frac{1}{2}\frac{1}{m(B_z)}\int_{B_z}\log\frac{1}{J_f(w)}\,dm(w)\right). \tag{5.2}$$

To prove the bi-Lipschitz property of HQC mappings in the plane with respect to the quasihyperbolic metric we will need the quasiconformal version of the Koebe-distortion theorem established by Astala and Gehring.

Theorem 5.1 ([19, Theorem 1.8]) *Suppose that D and D' are domains in $\mathbb{R}^n$. If $f : D \longrightarrow D'$ is K-quasiconformal, then*

$$\frac{1}{c}\frac{d(f(z), \partial D')}{d(z, \partial D)} \le \alpha_f(z) \le c\frac{d(f(z), \partial D')}{d(z, \partial D)}$$

for $z \in D$, where c is a constant which depends only on K and n.

Theorem 5.2 ([99]) *Suppose D and D' are proper subdomains of $\mathbb{R}^2$. If $f : D \longrightarrow D'$ is K-quasiconformal and harmonic, then it is bi-Lipschitz with respect to the quasihyperbolic metrics on D and D'.*

Proof Using the assumption that f is harmonic, we have locally the representation

$$f(z) = g(z) + \overline{h(z)},$$

where g and h are analytic functions. Then the Jacobian $J_f(z) = |g'(z)|^2 - |h'(z)|^2 > 0$ (note that $g'(z) \ne 0$).

Furthermore,

$$J_f(z) = |g'(z)|^2\left(1 - \frac{|h'(z)|^2}{|g'(z)|^2}\right) = |g'(z)|^2\left(1 - |\omega(z)|^2\right),$$

where $\omega(z) = \frac{h'(z)}{g'(z)}$ is analytic and $|\omega| < 1$. Now we have

$$\log\frac{1}{J_f(z)} = -2\log|g'(z)| - \log(1 - |\omega(z)|^2).$$

Note that the first term is a harmonic function. It is well known that the logarithm of the modulus of an analytic function is harmonic everywhere except where that analytic function vanishes, but $g'(z) \neq 0$ everywhere (see [124, p. 141]).

The second term can be expanded into the series

$$\sum_{k=1}^{\infty} \frac{|\omega(z)|^{2k}}{k},$$

and each term is subharmonic (note that ω is analytic).

So, $-\log(1 - |\omega(z)|^2)$ is a continuous function represented as a locally uniform sum of subharmonic functions. Thus it is also subharmonic.

It follows that

$$\log \frac{1}{J_f(z)} \text{ is a subharmonic function.} \tag{5.3}$$

It should be noted that the representation $f(z) = g(z) + \overline{h(z)}$ is local, but this suffices for our conclusion (5.3).

From (5.3), we have

$$\frac{1}{m(B_z)} \int_{B_z} \log \frac{1}{J_f(w)} dm(w) \geq \log \frac{1}{J_f(z)}.$$

Applying this with (5.2), we have

$$\frac{1}{\alpha_f(z)} \geq \exp\left(\frac{1}{2} \log \frac{1}{J_f(z)}\right) = \frac{1}{\sqrt{J_f(z)}},$$

and therefore

$$\sqrt{J_f(z)} \geq \alpha_f(z).$$

From the first inequality in Theorem 5.1 we have

$$\sqrt{J_f(z)} \geq \frac{1}{c} \frac{d(f(z), \partial D')}{d(z, \partial D)}. \tag{5.4}$$

Note that

$$J_f(z) = |g'(z)|^2 - |h'(z)|^2 \leq |g'(z)|^2$$

and by K-quasiconformality of f, $|h'| \leq k|g'|$, $0 \leq k < 1$, where $K = \frac{1+k}{1-k}$ (see (2.17)).

This gives $J_f \geq (1 - k^2)|g'|^2$. Hence

$$\sqrt{J_f} \asymp |g'| \asymp |g'| + |h'| = L(f, z),$$

where

$$L(f, z) = \max_{|h|=1} |f'(z)h|.$$

We finish by observing that (5.4) and the above asymptotic relation give us

$$L(f, z) \geq \frac{1}{c} \frac{d(f(z), \partial D')}{d(z, \partial D)}, \quad c = c(k).$$

To establish the reversed inequality, we again use $J_f(z) \geq (1 - k^2)|g'(z)|^2$, i.e.,

$$\sqrt{J_f(z)} \geq \sqrt{1 - k^2}|g'(z)|. \tag{5.5}$$

Furthermore, we know that for $n = 2$

$$\alpha_f(z) = \exp\left(\frac{1}{m(B_z)} \int_{B_z} \log\sqrt{J_f(x)}\, dm(w)\right).$$

Using (5.5)

$$\frac{1}{m(B_z)} \int_{B_z} \log\sqrt{J_f(x)}\, dm(w) \geq \frac{1}{m(B_z)} \int_{B_z} \log\sqrt{1 - k^2} + \log|g'(w)|\, dm(w)$$

$$= \log\sqrt{1 - k^2} + \frac{1}{m(B_z)} \int_{B_z} \log|g'(w)|\, dm(w)$$

$$= \log\sqrt{1 - k^2} + \log|g'(z)|.$$

Now we have the following by the harmonicity of $\log|g'|$

$$\alpha_f(z) = \exp\left(\frac{1}{m(B_z)} \int_{B_z} \log\sqrt{J_f(x)}\, dm(w)\right)$$

$$\geq \exp(\log\sqrt{1 - k^2} + \log|g'(z)|)$$

$$= \sqrt{1 - k^2}|g'(z)|$$

$$\geq \frac{1}{2}\sqrt{1 - k^2}(|g'(z)| + |h'(z)|)$$

$$= \frac{\sqrt{1 - k^2}}{2} L(f, z).$$

Applying the second inequality of Theorem 5.1, we get

$$L(f, z) \leq c\sqrt{J_f(z)} \leq c\alpha_f(z) \leq c\frac{d(f(z), \partial D')}{d(z, \partial D)}, \quad c = c(k).$$

Summarizing,

$$L(f,z) \asymp \frac{d(f(z), \partial D')}{d(z, \partial D)},$$

however, quasiconformality gives us

$$L(f, z) \asymp l(f, z),$$

where

$$l(f, z) = \min_{|h|=1} |f'(z)h|.$$

Therefore, we have

$$l(f, z) \asymp \frac{d(f(z), \partial D')}{d(z, \partial D)}.$$

This pointwise result, via integration along curves, easily gives

$$k_{D'}(f(z_1), f(z_2)) \asymp k_D(z_1, z_2).$$

This completes the proof.   □

5.2   When Part of Boundary Is Flat

In this section we give a method of achieving local bi-Lipschitz behavior when part of the boundary is flat. This is a local generalization of the work of Kalaj and Pavlović [79]. Our approach is to use the boundary Harnack inequality for this problem.

The following theorem will play an important role in our proofs.

Theorem 5.3 ([99]) *Let* $f : \Omega \longrightarrow \mathbb{C}$ *be a harmonic map whose Jacobian determinant* $J = |f_z|^2 - |f_{\bar{z}}|^2$ *is positive everywhere in* Ω*. Then* $\log J$ *is a superharmonic function.*

We note that this theorem has also been used in establishing the minimum principle for the Jacobian determinant, which is a novelty in the new analytic proof of the celebrated Radó–Kneser–Choquet theorem given by T. Iwaniec and J. Onninen [70].

We will also need the following boundary Harnack inequality ([47], exercise 6, p. 28).

Theorem 5.4 *Let u and v be positive harmonic functions on unit disk $\mathbb{D}$ in $\mathbb{R}^2$ with $u(0) = v(0)$ and let $I \subset \partial\mathbb{D}$ be an open arc and assume*

$$\lim_{z\to\zeta} u(z) = \lim_{z\to\zeta} v(z) = 0$$

for all $\zeta \in I$. Then for every compact $A \subset \mathbb{D} \cup I$ there is a constant $C(A)$ independent of u and v such that on $A \cap \mathbb{D}$

$$\frac{1}{C(A)} \leq \frac{u(z)}{v(z)} \leq C(A).$$

Proof We will consider the case $I = \partial\mathbb{D} \cap \mathbb{H}_-$ where $\mathbb{H}_- = \{z : Im(z) < 0\}$.

By our assumption u is positive and harmonic, so we have $u(z) = \int_{S^1} P_z(t)d\mu(t)$, where μ is a positive measure with $u(0) = \int_{S^1} d\mu = \mu(S^1)$. Applying the similar argument to v, we get that it is defined via a corresponding positive measure v.

Assume $v, u \geq 0$ are harmonic in $\mathbb{D}$ and $u\big|_I = v\big|_I = 0$, $u(0) = v(0) = 1$, i.e., $\mu(S^1) = v(S^1) = 1$. Since u is harmonic and μ is supported on $\{z : Im(z) \geq 0, |z| = 1\} = S^1_+$, we have

$$u(z) = \int_{S^1} \frac{1 - |z|^2}{|\xi - z|^2} d\mu(\xi) = \int_{S^1_+} \frac{1 - |z|^2}{|\xi - z|^2} d\mu(\xi).$$

For $\delta_0 = dist(A, supp(\mu))$ and $z \in A$ we have $dist(z, S^1_+) \geq \delta_0 > 0$ and

$$u(z) \leq (1 - |z|^2) \int_{S^1_+} \frac{1}{|\xi - z|^2} d\mu(\xi) \leq \frac{2(1 - |z|)}{\delta_0^2}.$$

Since $|\xi - z| \leq 2$ we have

$$v(z) \geq (1 - |z|) \int_{S^1_+} \frac{1}{|\xi - z|^2} dv(\xi) \geq \frac{1 - |z|}{4}$$

and we conclude that for $z \in A$ we have $u(z)/v(z) \leq \frac{8}{\delta_0^2}$ and analogously $v(z)/u(z) \leq \frac{8}{\delta_0^2}$, and hence

$$\frac{\delta_0^2}{8} \leq \frac{u(z)}{v(z)} \leq \frac{8}{\delta_0^2}.$$

$\square$

Using Theorem 5.4 we obtain the following special case of this result.

Theorem 5.5 ([103]) *Suppose that* $\mathbb{D}$ *is the unit disk and* $\mathbb{H}^2$ *is the upper-half plane in* $\mathbb{R}^2$. *If* $f : \mathbb{D} \longrightarrow \mathbb{H}^2$ *is HQC homeomorphism with* $f(\partial(\mathbb{D}_-)) \subseteq \mathbb{R}$, *then* $f|\mathbb{D}_-$ *is bi-Lipschitz with respect to the Euclidean metric, where* $\mathbb{D}_- = \{z : z \in \mathbb{D}, Im(z) < 0\}$.

Proof It is clear that we loss no generality by assuming that $f(0) = i$. We take the Möbius transformation

$$M(z) = \frac{1 - iz}{z - i}$$

such that $M(\pm 1) = \pm 1$, $M(0) = i$, $M(-i) = 0$ and choose

$$u = Im(f), \quad v = Im(M(z)) = \frac{1 - |z|^2}{|z - i|^2}.$$

Note that $u(0) = v(0) = 1$ and that for each ξ, such that $Im(\xi) < 0, |\xi| = 1$

$$\lim_{z \to \xi} u(z) = \lim_{z \to \xi} v(z) = 0.$$

Note that in our setting $Im(f(z)) \equiv d(f(z), \partial\mathbb{H}^2)$, so applying Theorem 5.4 we now have

$$\frac{1}{C(A)} \le \frac{d(f(z), \partial\mathbb{H}^2)}{\frac{1-|z|^2}{|z-i|^2}} \le C(A)$$

on $A \cap \mathbb{D}$ for some constant $C(A)$ and for every compact $A \subset \mathbb{D} \cup I$, where $I = \partial\mathbb{D} \cap \mathbb{H}_-$.
   Since $|z - i|^2 \le 4, d(z, \partial\mathbb{D}) = 1 - |z|$, it follows that

$$\frac{1}{2C(A)} \le \frac{d(f(z), \partial\mathbb{H}^2)}{d(z, \partial\mathbb{D})} \le 4C(A).$$

Using Theorem 5.1 we conclude that

$$\frac{1}{c} \le \alpha_f(x) \le c,$$

where c is a constant which depends only on A.
   We finish by noting that from the proof of the main theorem in [99] it follows that

$$\alpha_f(x) \asymp \|f'(x)\|,$$

and since f is quasiconformal, it follows that it is bi-Lipschitz. $\qquad\qquad\square$

We give a local version of Theorem 5.5.

Lemma 5.6 ([103]) *Let $D_+ = \mathbb{D} \cap \mathbb{H}^2$ and let $g : \mathbb{D} \to D_+$ be a harmonic K-quasiconformal mapping with*

$$g(\pm 1) = \pm 1, \qquad g(-i) = 0.$$

If $A \subset \overline{D}$ is a compact subset with $\delta_0 := \text{dist}(A, S^1 \cap \mathbb{H}^2) > 0$, then

$$\frac{1}{c(K, \delta_0)} \leq \frac{\text{dist}(g(z), \partial D_+)}{1 - |z|} \leq c(K, \delta_0), \qquad z \in A.$$

The constant $c(K, \delta_0) < \infty$ depends only on K and δ_0.

Proof First note that the map $g : \mathbb{D} \to D_+$ is η-quasisymmetric, where η depends only on K. To see this note that every K-quasiconformal mapping of the unit disk $\mathbb{D}$ fixing ± 1 and $-i$ is η-quasisymmetric, and so the case of our mapping is reduced to this fact using, for example, an appropriate bi-Lipschitz mapping from D_+ to $\mathbb{D}$.

Note that if $u(z) = Im(g(z))$, $z \in A$, then we have $c(K) \leq u(0) \leq 1$, and that,

$$\frac{1}{c(K, \delta_0)} \leq \frac{u(z)}{\text{dist}(g(z), \partial D_+)} \leq c(K, \delta_0)$$

for some constant $c(K) < \infty$ depending only on K and δ_0. Hence, we can argue similarly as in Theorem 5.5 to finish the proof. $\square$

By further developing the above ideas, we can consider local questions of bi-Lipschitz property phenomena when only part of the boundary is flat. In this case we need to use some quasiconformal geometry.

Definition 5.7 We say that $\partial\Omega$ is *flat* at some $x_0 \in \partial\Omega$ if, up to a rotation around the point x_0, we have

$$\partial\Omega \cap B^2(x_0, \rho) = [x_0 - \rho, x_0 + \rho]$$

for some $\rho > 0$.

Theorem 5.8 ([103]) *Suppose that $\mathbb{D}$ is the unit disk, Ω is simply connected, and $f : \mathbb{D} \longrightarrow \Omega$ is a harmonic and quasiconformal mapping such that $f(\mathbb{D}) = \Omega$. Suppose also that $\partial\Omega$ is flat at x_0, and that f is normalized so that $f(\pm 1) = x_0 \pm \rho$ with $f(-i) = x_0$.*
If $\Omega_1 = f^{-1}(B^2(x_0, \rho/2) \cap \Omega)$, then $f : \Omega_1 \longrightarrow B^2(x_0, \rho/2) \cap \Omega$ is bi-Lipschitz. Indeed,

$$\frac{1}{L_0} \leq \frac{|f(x) - f(y)|}{\rho|x - y|} \leq L_0$$

for some L_0 depending only on $K(f)$.

Proof By Lemma 5.6 for the conformal mapping $\phi : \mathbb{D} \to \tilde{\Omega} = f^{-1}[B^2(x_0, \rho) \cap \Omega]$ with $\phi(\pm 1) = \pm 1$ and $\phi(-i) = -i$, the statement follows as it is enough to notice that ϕ is bi-Lipschitz on $A = f^{-1}[B^2(x_0, \rho/2) \cap \Omega]$. $\qquad\square$

5.3 Bi-Lipschitz Property of HQC Mappings in Higher Dimensions

In this section we study counterparts in higher dimensions of the following result of M. Pavlović [123].

Theorem 5.9 ([123]) *Let u be a harmonic homeomorphism of $\mathbb{D}$ that extends continuously to the boundary, and $f : \mathbb{S}^1 \to \mathbb{S}^1$ be its boundary function. Then the following conditions are equivalent:*

1. *u is quasiconformal;*
2. *u is bi-Lipschitz in the Euclidean metric;*
3. *f is bi-Lipschitz and the Hilbert transformation of its derivative is in L^∞.*

In [121], Partyka and Sakan gave explicit estimations of the bi-Lipschitz constants for u expressed by means of the maximal dilatation K of u and $|u^{-1}(0)|$. Additionally, if $u(0) = 0$, they have asymptotically sharp estimates as $K \to 1$.

In order to prove Theorem 5.9 Pavlović had to make a deep and detailed analysis of the boundary values of f. By analyzing these boundary values, he achieved the Lipschitz-property for every harmonic quasiconformal mapping of the disk. In higher dimensions, Pavlović's approach seems difficult to generalize. Instead it would seem possible that the Lipschitz-property follows by the regularity theory of elliptic PDEs. Indeed, such an approach was taken by Kalaj [76] but the proof in [76] is quite long and technical. A simple and self-contained argument showing the Lipschitz property in all dimensions is given in [21].

Recall that $L^p(\Omega, \mathbb{R}^n)$ for $p \in [1, \infty)$ is the Banach space of all measurable functions $f : \Omega \longrightarrow \mathbb{R}^n$ such that

$$\int_\Omega |f|^p < \infty$$

with the norm defined as

$$\|f\|_p = \left(\int_\Omega |f|^p \right)^{\frac{1}{p}}.$$

Recall that L^∞ is the Banach space of all measurable functions $f : \Omega \longrightarrow \mathbb{R}^n$ such that there is a constant $c \in [0, \infty)$ such that

$$|f| \le c, \quad \text{a.e. in } \Omega$$

with the norm defined as

$$\|f\|_\infty = \inf\{c \geq 0 \; : \; |f| \leq c \text{ a.e. in } \Omega\}.$$

Recall also that $L_{loc}^p(\Omega, \mathbb{R}^n)$ for $p \in [1, \infty]$ is the vector space of all measurable functions $f : \Omega \longrightarrow \mathbb{R}^n$ such that for every compact set $K \subseteq \Omega$ holds $f|_K \in L^p(K)$. Instead of $L^p(\Omega, \mathbb{R})$ we write $L^p(\Omega)$.

Definition 5.10 ([65, Definition A.13, p. 148]) Let $\Omega \subseteq \mathbb{R}^n$ be open and $u \in L_{loc}^1(\Omega)$. A function $v \in L_{loc}^1$ is called a *weak derivative* of u if

$$\int_\Omega \varphi(x) v(x)\, dx = - \int_\Omega u(x) \nabla \varphi(x)\, dx$$

for every $\varphi \in C_C^\infty(\Omega)$. We denote such a function v by Du. For $1 \leq p \leq \infty$ we define the Sobolev space

$$W^{1,p}(\Omega) = \{u \in L^p(\Omega) \; : \; Du \in L^p(\Omega, \mathbb{R}^n)\}$$

and we define the norm

$$\|u\|_{W^{1,p}(\Omega)} = \left(\int_\Omega |u|^p + \int_\Omega |Du|^p \right)^{\frac{1}{p}}.$$

For harmonic K-quasiconformal mappings $f = (f^1, \ldots, f^n) : \mathbb{B}^n \to \mathbb{B}^n$ by harmonicity we have

$$\Delta (f^j)^2(x) = 2|\nabla f^j(x)|^2, \qquad j = 1, \ldots, n.$$

We say that a mapping f defined in a domain $\Omega \subset \mathbb{R}^n$ has the co-Lipschitz property with constant $L \geq 1$, if

$$|f(x) - f(y)| \geq \frac{1}{L}|x - y| \qquad \forall\, x, y \in \Omega. \tag{5.6}$$

Naturally, for mappings in (5.6) the Jacobians are nonvanishing everywhere.

Lemma 5.11 (Wood [42, p. 26]) *In dimensions $n \geq 3$, the Jacobian of a harmonic homeomorphism may vanish.*

Example 5.12 Let $f : \mathbb{R}^3 \longrightarrow \mathbb{R}^3$ be as follows:

$$f(x, y, z) = (u, v, w),$$

where

$$u = x^3 - 3xz^2 + yz, \quad v = y - 3xz, \quad w = z.$$

It is clear that f is harmonic. To see that f is univalent suppose that

$$f(x_1, y_1, z_1) = f(x_2, y_2, z_2) = (u, v, w).$$

Then $w = z_1 = z_2$ and

$$v = y_1 - 3x_1 w = y_2 - 3x_2 w,$$

which implies

$$u = x_1^3 + w(y_1 - 3x_1 w) = x_2^3 + w(y_2 - 3x_2 w).$$

It follows that

$$x_1^3 + vw = x_2^3 + vw,$$

so $x_1 = x_2$ and $y_1 = y_2$. The calculations show that the mapping $(u, v, w) \mapsto (x, y, z)$ is defined by

$$x = (u - vw)^{1/3}, \quad y = v + 3w(u - vw)^{1/3}, \quad z = w,$$

is an inverse for f. Thus, f is a (univalent) harmonic mapping of $\mathbb{R}^3$ onto $\mathbb{R}^3$. A straightforward calculation reveals that f has the Jacobian

$$J_f(x, y, z) = 3x^2,$$

which vanishes on the plane $x = 0$.

Note that the main difficulty here is in finding lower bounds for $|f(x) - f(y)|$ in terms of the distance between x and y. However, in general dimensions it is not known if harmonic quasiconformal mappings of the ball have nonvanishing Jacobian. Note now that, by the Lewy theorem [95], if the gradient $f = \nabla u$ of a (real valued) harmonic function defines a homeomorphism $f : \Omega \to \Omega'$ where $\Omega, \Omega' \subset \mathbb{R}^3$, then the Jacobian J_f does not vanish. Moreover, $J_f = \mathscr{H}_u$, where $\mathscr{H}_u$ denotes the Hessian of u, and here we have the theorem of Gleason and Wolff [56, Theorem A] which shows that in dimension three, the function $\log \det (\mathscr{H}_u)$ is superharmonic outside the zeroes of the Hessian. The following result collects these facts.

Theorem 5.13 (Lewy–Gleason–Wolff) *Suppose* $u : \Omega \to \mathbb{R}$ *is a harmonic function, such that* $f(x) := \nabla u(x)$ *defines a homeomorphism between the domains* Ω *and* $\Omega' \subset \mathbb{R}^3$. *Then*

$$\log J_f(z) = \log \det (\mathscr{H}_u) \text{ is superharmonic in } \Omega. \tag{5.7}$$

The usual Sobolev embedding theorem in whole space is given in [93, Theorem 1, p. 263]. In the paper [21], the local form of Sobolev embedding for the unit ball is proved using Green's function and the estimates on Riesz potential.

Lemma 5.14 ([21]) *Suppose that* $w \in C^2(\mathbb{B}^n) \cap C(\overline{\mathbb{B}^n})$, *that* $h \in L^p(\mathbb{B}^n)$ *for some* $1 < p < \infty$ *and*

$$\Delta w = h \ \text{in} \ \mathbb{B}^n, \ \text{with} \ w\big|_{S^{n-1}} = 0,$$

(a) If $1 < p < n$, *then*

$$\|\nabla w\|_{L^q(\mathbb{B}^n)} \leq c(p,n)\|h\|_{L^p(\mathbb{B}^n)}, \qquad \frac{1}{q} = \frac{1}{p} - \frac{1}{n}.$$

(b) If $n < p < \infty$, *then*

$$\|\nabla w\|_{L^\infty(\mathbb{B}^n)} \leq c(p,n)\|h\|_{L^p(\mathbb{B}^n)}.$$

Proof The function w can be represented in terms of the Green function $G_{\mathbb{B}^n}(x,y)$ of the unit ball:

$$w(x) = \int_{\mathbb{B}^n} G_{\mathbb{B}^n}(x,y)h(y)dm(y), \qquad x \in \mathbb{B}^n.$$

The Green function and its gradient

$$\nabla_x G_{\mathbb{B}^n}(x,y) = c_1(n)\left[\frac{x-y}{|x-y|^n} + |y|^n \frac{y-|y|^2 x}{|y-|y|^2 x|^n}\right]$$

can be explicitly calculated. Details of calculation can be found in [93, p. 40]. Since $|y||x-y| \leq |y-|y|^2 x|$ for all $x,y \in \mathbb{B}^n$, the gradient is bounded by

$$|\nabla_x G_{\mathbb{B}^n}(x,y)| \leq 2c_1(n)|y-x|^{1-n} \quad \text{for} \quad x,y \in \mathbb{B}^n.$$

Therefore $\|\nabla w\|_{L^q(\mathbb{B}^n)} \leq c\|\mathscr{I}_1 h\|_{L^q(\mathbb{R}^n)}$, where $\mathscr{I}_s h$ denotes the Riesz potential of order s. It follows that Lemma 5.14(*a*) reduces to the well-known boundedness properties of the Riesz potentials,

$$\|\mathscr{I}_s h\|_{L^q(\mathbb{R}^n)} \leq c(s,p,q)\|h\|_{L^p(\mathbb{R}^n)}, \qquad \frac{1}{q} = \frac{1}{p} - \frac{s}{n},$$

given, e.g., in [150, p. 119]. We note that the bound in *b*) is easier and it follows from Hölder's inequality, since $x \mapsto |x-y|^{1-n} \in L^q(\mathbb{B}^n)$ for every $1 \leq q < \frac{n}{n-1}$. □

Lemma 5.14 leads us to a quick proof for the following result.

Corollary 5.15 ([21, Lemma 2.1]) *Suppose $w \in W^{2,1}_{loc}(\mathbb{B}^n) \cap C(\overline{\mathbb{B}^n})$, $n \geq 2$, is such that*

$$w\big|_{S^{n-1}} = 0, \quad \text{with} \quad \int_{\mathbb{B}^n} |\nabla w|^{p_0}\, dm < \infty \text{ for some } n < p_0 < \infty. \tag{5.8}$$

If w satisfies the following uniform differential inequality,

$$|\Delta w(x)| \leq a|\nabla w(x)|^2 + b, \qquad x \in \mathbb{B}^n, \tag{5.9}$$

for some constants $a, b < \infty$, we then have

$$\|\nabla w\|_{L^\infty(\mathbb{B}^n)} \leq M < \infty, \tag{5.10}$$

where $M = M(a, b, p_0, n, \|\nabla w\|_{p_0})$. In particular, w is Lipschitz continuous.

Proof Applying (5.9), we have

$$\Delta w(x) = h(x), \qquad \text{for } x \in \mathbb{B}^n, \tag{5.11}$$

where

$$h(x) = c(x)\left(|\nabla w(x)|^2 + 1\right) \tag{5.12}$$

and $\|c\|_\infty \leq \max\{a, b\}$; one can simply define

$$c(x) := \Delta w(x)\left(|\nabla w(x)|^2 + 1\right)^{-1}$$

for almost every $x \in \mathbb{B}^n$.

Note that here, by our assumptions, $\nabla w \in L^{p_0}(\mathbb{B}^n)$, but with Sobolev embedding one can improve this integrability up to

$$\nabla w \in L^s(\mathbb{B}^n) \qquad \text{where } s > 2n. \tag{5.13}$$

Namely, if $p_0/2 < n < p_0$, then $h(x) = c(x)\left(|\nabla w(x)|^2 + 1\right) \in L^{p_0/2}(\mathbb{B}^n)$ and (5.11) with Lemma 5.14(a) give us

$$\nabla w \in L^{p_1}(\mathbb{B}^n), \qquad p_1 = \frac{p_0 n}{2n - p_0} > p_0, \tag{5.14}$$

which is a strict improvement in the integrability.

Note that if initially $p_0 = n(1 + \varepsilon)$, $\varepsilon > 0$, then

$$p_1 = \frac{p_0 n}{2n - p_0} = n\frac{1 + \varepsilon}{1 - \varepsilon} > n(1 + 2\varepsilon).$$

So one can iterate this feedback argument, getting $\nabla w \in L^{p_\ell}(\mathbb{B}^n)$, $\ell = 0, 1, 2, \ldots$ with $p_\ell > n(1 + 2^\ell \varepsilon)$, until the condition (5.13) is reached. (If it turns out that for some exponent $p_\ell = 2n$, we can choose p_0 a little smaller so that this degeneracy does not happen.)

Once (5.13) is achieved, (5.11)–(5.12) with Sobolev embedding and Lemma 5.14(b) give $\|\nabla w\|_\infty < \infty$. Note that the proof also gives a bound for $\|\nabla w\|_\infty$ that depends only on the constants a and b, the exponent p_0, and the initial norm $\|h\|_{p_0/2} \le \max\{a, b\}(\|\nabla w\|^2_{p_0} + 1)$. □

It should be noted that the above iteration argument fails if in (5.8) one assumes integrability only for some $1 \le p_0 < n$. Thus, the following Gehring's theorem [50] in case of quasiconformal mappings becomes particularly useful here.

Lemma 5.16 ([50]) *For every quasiconformal mapping* $f : \mathbb{R}^n \to \mathbb{R}^n$ *we have the following higher integrability property*

$$\int_{\mathbb{B}^n} |Df(x)|^p dm \le C < \infty, \qquad p = p(n, K) > n, \tag{5.15}$$

where for mappings of the whole space $\mathbb{R}^n$, *the constant* C *depends only on* n *and distortion* $K(f)$.

Theorem 5.17 ([21]) *If* $n \ge 2$ *and* $f : \mathbb{B}^n \to \mathbb{B}^n$ *is a harmonic and* K-*quasiconformal mapping, then*

$$|f(x) - f(y)| \le L|x - y|, \qquad x, y \in \mathbb{B}^n,$$

where L *depends only on the distortion* K, *dimension* n, *and* $\mathrm{dist}(f(0), S^{n-1})$.

Proof In case $f : \mathbb{B}^n \to \mathbb{B}^n$ is K-quasiconformal, we can compose it with a Möbius transform ψ preserving the ball, such that $f \circ \psi(0) = 0$. Using Schwarz reflection one can then extend $f \circ \psi$ to $\mathbb{R}^n$ and apply (5.15) to this mapping. Unfolding the Möbius transformation after a change of variables, we see that each K-quasiconformal mapping $f : \mathbb{B}^n \to \mathbb{B}^n$ satisfies (5.15) with $C = C(n, K, \mathrm{dist}(f^{-1}(0), S^{n-1}))$

Moreover, if f is harmonic, consider the function

$$w(x) = 1 - |f(x)|^2, \qquad x \in \mathbb{B}^n.$$

Recall that quasiconformal mappings of $\mathbb{B}^n$ extend continuously to the boundary (see [155, Theorem 17.20, p. 61]), so we have that $w(x)$ satisfies the assumptions of Corollary 5.15. To check the condition (5.9) note that $w = u \circ f$ where

$$u(x) = 1 - |x|^2 \quad \text{with} \quad \nabla u(x) = -2x, \qquad x \in \mathbb{B}^n.$$

Hence $\nabla w(x) = Df^t(x)\nabla u(f(x))$ satisfies

$$\frac{2}{K}|f(x)| |Df(x)| \le |\nabla w(x)| \le 2|f(x)| |Df(x)| \tag{5.16}$$

with

$$|\Delta w(x)| = 2||Df(x)||^2 \leq 2n^2|Df(x)|^2, \qquad x \in \mathbb{B}^n,$$

where $||Df(x)||^2$ denotes the Hilbert–Schmidt norm of the differential matrix.

We note that the above argument already establishes (5.9) but to see the explicit dependence of a and b on properties of the mapping f we first need to note that there is a constant $\delta = \delta(n, K, a, \text{dist}(f(0), S^{n-1}))$ such that

$$1 - |x| + |f(x)| \geq \delta > 0, \qquad \text{for all } x \in \mathbb{B}^n. \tag{5.17}$$

To see this note that quasiconformal mappings of $\mathbb{B}^n$ are quasi-isometries in the hyperbolic metric [158]. Hence either $h_{\mathbb{B}^n}(f(x), f(0)) \geq 1 + h_{\mathbb{B}^n}(f(0), 0)$ (which implies $h_{\mathbb{B}^n}(f(x), 0) \geq 1$) and $|f(x)| \geq \frac{e-1}{e+1}$ hold, or else we have

$$h_{\mathbb{B}^n}(x, 0) \leq c(K)(1 + h_{\mathbb{B}^n}(f(x), f(0))) \leqslant c(K)(2 + h_{\mathbb{B}^n}(f(0), 0)).$$

In the latter case $1 - |x| \geqslant e^{-M}$, where $M = c(K)(2 + h_{\mathbb{B}^n}(f(0), 0))$. Hence (5.17) holds, and we have

$$|\Delta w| \leq \frac{4n^2}{\delta^2}[(1 - |x|)^2 + |f(x)|^2]|Df(x)|^2$$
$$\leq \frac{2Kn^2}{\delta^2}|\nabla w|^2 + \frac{4n^2}{\delta^2}(1 - |x|)^2|Df(x)|^2.$$

The last term is controlled by basic ellipticity bounds [55, p.38], or more precisely, the Bloch norm bounds

$$(1 - |x|)|Df(x)| \leq c(n)\|f\|_\infty \tag{5.18}$$

which is valid for every harmonic function. Hence, (5.9) holds with $a = K^2n^2\delta^{-2}$, $b = 4n^2c(n)^2\delta^{-2}$, so that $\nabla w \in L^\infty(\mathbb{B}^n)$ by Corollary 5.15. In a similar manner (5.16)–(5.18) give us

$$|Df(x)| \leqslant \frac{1}{\delta}(1 - |x| + |f(x)|)|Df(x)| \leqslant \frac{c(n)}{\delta} + \frac{K\|\nabla w\|_\infty}{2\delta}.$$

Thus f is a Lipschitz mapping, with an explicit bound

$$L = L(n, K, dist(f(0), S^{n-1}))$$

for its Lipschitz constant. $\qquad\qquad\qquad\qquad\qquad\qquad\qquad\qquad\qquad\qquad\qquad\quad \square$

Definition 5.18 (Muckenhoupt Classes) Let w be a locally integrable nonnegative function in $\mathbb{R}^n$ such that $0 < w < \infty$ almost everywhere. Then w belongs to the

Muckenhoupt class A_p for $1 < p < \infty$, or that is an A_p-*weight* whenever there is a constant $c_{p,w}$ such that

$$\frac{1}{|B|} \int_B w \, dx \leq c_{p,w} \left(\frac{1}{|B|} \int_B w^{\frac{1}{1-p}} \, dx \right)^{1-p}$$

for all balls B in $\mathbb{R}^n$. We let w belong to A_1, or that w is an A_1-*weight* if there is a constant $c_{1,w} \geq 1$ such that

$$\frac{1}{|B|} \int_B w \, dx \leq c_{1,w} \operatorname{ess\,inf}_B w$$

for all balls B in $\mathbb{R}^n$. The union of all Muckenhoupt classes A_p is denoted by A_∞, or more precisely,

$$A_\infty = \bigcup_{p>1} A_p.$$

We shall say that w is an A_∞-*weight* if it belongs to A_∞, i.e., if w is an A_p-weight for some $p > 1$.

Recall that for a quasiconformal mapping, the Jacobian J_f is an A_∞-weight (see [64, Theorem 15.32]), $\alpha_f(z)$ is comparable to $\left(\frac{1}{m(B_z)} \int_{B_z} J_f^p \right)^{1/p}$ for every $0 < p \leq 1$, and hence we could have used such averages as well. From the other side, in the case $n = 2$ and f conformal, we have

$$\alpha_f(z) = |f'(z)|,$$

and therefore the choice (5.1) above appears to be a natural one.

We now need a version of the classical Hopf lemma.

Lemma 5.19 (Hopf Lemma [23, Exercise I-25]) *Suppose that u is real valued, nonconstant, and harmonic on $\overline{\mathbb{B}}^n$. Show that if u attains its maximum value on $\overline{\mathbb{B}}^n$ at $\zeta \in S$, then there is a positive constant c such that*

$$u(\zeta) - u(r\zeta) \geq c(1 - r)$$

for all $r \in (0, 1)$. Conclude that $(D_n u)(\zeta) > 0$.

It should be noted that this lemma is frequently applied in various contexts for estimating superharmonic functions in terms of boundary distance. For example, in [116] M. Mateljević used this well-known approach for harmonic quasiconformal mappings.

Theorem 5.20 ([21]) *Suppose $f : \mathbb{B}^n \longrightarrow \Omega$ is a harmonic quasiconformal mapping, with $\Omega \subset \mathbb{R}^n$ a convex subdomain. Then*

$$\alpha_f(x) \geq c_0\, d(f(0), \partial\Omega) > 0, \qquad x \in \mathbb{B}^n, \tag{5.19}$$

where the constant $c_0 = c_0(n, K)$ depends only on the dimension n and distortion $K = K(f)$.

Proof For $z \in \mathbb{B}^n$, we have

$$d(f(z), \partial\Omega) = \inf_p d(f(z), p),$$

where the infimum is taken over all lines p outside the domain. Note that the function

$$h_p(z) = d(f(z), p)$$

is positive and harmonic in $\mathbb{B}^n$. Applying the usual Harnack inequality in $\mathbb{B}^n$ to each h_p, we get,

$$h_p(z) \geq \frac{1 - |z|}{(1 + |z|)^{n-1}}\, h_p(0).$$

Since $d(f(0), p) \geq d(f(0), \partial\Omega)$, we have

$$h_p(z) \geq \frac{1 - |z|}{(1 + |z|)^{n-1}}\, d(f(0), \partial\Omega).$$

Taking the infimum of the $h_p(z)$ over all p gives us

$$d(f(z), \partial\Omega) \geq \frac{1 - |z|}{(1 + |z|)^{n-1}}\, d(f(0), \partial\Omega).$$

Note that since

$$d(z, \partial\mathbb{B}^n) = 1 - |z|,$$

the last inequality can be rewritten as follows:

$$\frac{d(f(z), \partial\Omega)}{d(z, \partial\mathbb{B}^n)} \geq \frac{d(f(0), \partial\Omega)}{(1 + |z|)^{n-1}}.$$

Applying Theorem 5.1 and quasiconformality of f, we conclude that

$$\alpha_f(z) \geq c(n, K) d(f(0), \partial\Omega).$$

$\square$

It follows that the co-Lipschitz property can be established if the usual derivative can be estimated from below by the average derivative. We note that in dimension $n = 2$ this can be done using the main result of the paper [99].

Theorem 5.21 *Suppose* $\Omega, \Omega' \subset \mathbb{R}^2$ *are planar domains and* $f : \Omega \to \Omega'$ *a harmonic quasiconformal mapping. Then* $\log J_f$ *is superharmonic in* Ω.

Using the superharmonicity of $\log J_f$ for the harmonic quasiconformal mapping f defined in the unit disk $\mathbb{B}^2$, we get

$$\log |Df(x)|^2 \geq \log J_f(z) \geq \frac{1}{m(B_z)} \int_{B_z} \log J_f \, dm = \log \alpha_f(z)^2, \qquad z \in \mathbb{B}^2.$$
(5.20)

Combining this estimate with Theorem 5.20 proves that for every harmonic quasiconformal mapping from the disk onto a convex domain the lower bound

$$\inf_{|h|=1} |Df(x)h| \geq |Df(x)|/K \geq \alpha_f(x)/K \geq cd(f(0), \partial\Omega)$$
(5.21)

holds for some constant $c > 0$. It follows from this that f is co-Lipschitz. This is a new proof of theorem [73, Th 3.5]. In fact we have here the following fact.

Corollary 5.22 ([21]) *Suppose* $\Omega, \Omega' \subset \mathbb{R}^2$ *are simply connected domains and* $f : \Omega \to \Omega'$ *is a harmonic quasiconformal mapping. If* Ω' *is convex and the Riemann map of* Ω *has derivative bounded from above, then* f *has the co-Lipschitz property (5.6).*

The proof relies on applying (5.21) to $f \circ g$, where $g : D \to \Omega$ is the Riemann map. Hence, in particular, in Corollary 5.22 the boundary of Ω need not be C^1 or even Lipschitz. For example, $g(z) = 2z - z^2$ is a conformal map from D onto a cardioid, with cusp at $1 = g(1)$.

In a similar manner, combining (5.20) with Theorems 5.17 and 5.20, we have a new proof for Pavlović's theorem for $\mathbb{B}^2$ (see Theorem 5.9).

Finally, to prove the higher dimensional version of the Pavlović theorem, we will use an argument analogous to (5.20) and Theorem 5.21.

Theorem 5.23 ([21]) *Suppose* $f : \mathbb{B}^3 \to \mathbb{B}^3$ *is a harmonic quasiconformal mapping, which is also a gradient mapping, that is* $f = \nabla u$ *for some function* u *harmonic in the unit ball* $\mathbb{B}^3$, *then* f *is a bi-Lipschitz mapping.*

Proof Note that if $f = \nabla u$ is a quasiconformal harmonic gradient mapping in $\mathbb{B}^3$, then as in (5.20), using Theorem 5.13 we conclude that

$$\alpha_f(x)^3 \leq J_f(x) \leq K(f)^2 \inf_{|h|=1} |Df(x)h|^3.$$

Hence, when $f(\mathbb{B}^3) = \mathbb{B}^3$, or more generally when the target domain is convex, Theorem 5.20 gives us that f is co-Lipschitz. The Lipschitz-property follow from Theorem 5.17. This finishes the proof.   □

The above approach has a few consequences also in more general subdomains of $\mathbb{R}^n$. We have a generalization of a theorem 5.2.

Theorem 5.24 ([21]) *Consider domains* $\Omega, \Omega' \subset \mathbb{R}^3$, *and let* $f : \Omega \to \Omega'$ *be a harmonic quasiconformal homeomorphism which is also a gradient mapping, say* $f = \nabla u$ *for some function u harmonic in* Ω. *Then* f *is bi-Lipschitz with respect to the corresponding quasihyperbolic metrics,*

$$\frac{1}{M} k_\Omega(x, y) \le k_{\Omega'}(f(x), f(y)) \le M k_\Omega(x, y), \qquad x, y \in \Omega,$$

where the constant M depends only on the distortion $K(f)$.

Proof Using (5.7) and Theorem 5.13, we get that $\alpha_f(x)^3 \le J_f(x)$. Looking from the other side, $Df(x)$ is a vector valued harmonic function, whose norm is subharmonic and therefore

$$\|Df(x)\|^3 \le \frac{1}{m(B_x)} \int_{B_x} \|Df\|^3 \, dm \le \frac{K}{m(B_x)} \int_{B_x} J_f \, dm$$

$$\le C(K, n) \exp[\frac{1}{m(B_x)} \int_{B_x} \log J_f \, dm] = C(K, n)\alpha_f(x)^3,$$

where the last inequality follows from the fact that J_f is an A_∞-Muckenhoupt weight. Thus $\alpha_f(x) \simeq \inf_{|h|=1} |Df(x)h| \simeq \sup_{|h|=1} |Df(x)h|$, and the conclusion follows as in [99]. □

The proof of Theorem 5.23 gives immediately the following consequence.

Corollary 5.25 ([21]) *Suppose* $f : \mathbb{B}^3 \to \Omega$ *is quasiconformal. If* Ω *is convex and* $f = \nabla u$ *is the gradient of a harmonic function, then* f *has the co-Lipschitz property (5.6).*

We note that the method of the proof of Theorem 5.17 works also for more general domains. We also note that we have the following result of Kalaj [76] as consequence.

Corollary 5.26 ([21]) *If* $f : \mathbb{B}^3 \to \Omega$ *is a harmonic quasiconformal mapping, where* $\Omega \subset \mathbb{R}^n$ *is a domain with* C^2-*boundary, then* f *is a Lipschitz mapping.*

Proof Consider this time $w(x) = \text{dist}(f(x), \partial\Omega)$ near $\partial\Omega$, and choose some smooth extension to Ω. Then w satisfies the inequality (5.9) (see [76]), so $\|\nabla w\|_\infty < \infty$ by Corollary 5.15, and we obtain the Lipschitz bounds for f as in the proof of Theorem 5.17. □

Finally we mention that combining the above results, we have the following theorem as consequence.

Theorem 5.27 ([21]) *Suppose* Ω *is a convex subdomain of* $\mathbb{R}^3$ *with* C^2-*boundary, and let* $f : \mathbb{B}^3 \to \Omega$ *be a harmonic quasiconformal homeomorphism. If* $f = \nabla u$ *is a harmonic gradient mapping, then* f *is bi-Lipschitz with respect to the Euclidean metric.*

Chapter 6
Quasi-Nearly Subharmonic Functions and QC Mappings

Let G be a domain in $\mathbb{R}^n$, $f : G \to \mathbb{R}^n$ a harmonic map, and $\mathscr{A}$ a class of self-homeomorphisms of G. We study in this chapter what can be said about the functions of the form $f \circ h, h \in \mathscr{A}$. For example, we show that if $n = 2$ and $\mathscr{A}$ is the class of conformal maps, then the functions in this class are also harmonic. However, if $\mathscr{A}$ is the class of harmonic maps, or quasiconformal harmonic maps, this statement is no longer true. Our presentation here is based on [38, 87, 90] and the survey paper [153] which has a slightly different focus.

6.1 Quasi-Nearly Subharmonic Functions and Conformal Mappings

Let $B^2(z, r)$ denote the Euclidean disk with center z and radius r, and let m be the Lebesgue measure in $\mathbb{C}$ normalized so that the measure of the unit disk equals 1. Then if $u : \Omega \longrightarrow \mathbb{R}$ is a subharmonic nonnegative function on a domain $\Omega \subset \mathbb{C}$ and $p \geqslant 1$, then

$$u(z)^p \leqslant \frac{1}{r^2} \int_{B^2(z,r)} u^p \, dm. \tag{6.1}$$

If $0 < p < 1$, then (6.1) need not hold, but there is a constant $C = C(p) \geqslant 1$ such that

$$u(z)^p \leqslant \frac{C}{r^2} \int_{B^2(z,r)} u^p \, dm. \tag{6.2}$$

This fact, essentially due to Hardy and Littlewood [57], was first proved by Fefferman and Stein [44, Lemma 2]. The proof of Fefferman and Stein is reproduced by Garnett [46, Lemma 3.7]. It should be noted that Fefferman and Stein originally

© Springer Nature Switzerland AG 2019

V. Todorčević, *Harmonic Quasiconformal Mappings and Hyperbolic Type Metrics*,
https://doi.org/10.1007/978-3-030-22591-9_6

considered only the case when $u = |v|$ and v is harmonic, but their proof applies also in the case of nonnegative subharmonic functions. For proofs of (6.2) see [134], [122, Theorem 1], [135, Lemma 2.1], [38, 125, 136].

That the validity of (6.2) for some p implies its validity for all p was observed in [3], [122, Theorem 1], and [135, Lemma 2.1].

Following [125], we call a Borel measurable function $u : \Omega \longrightarrow [0, \infty]$ on a subdomain Ω of the complex plane $\mathbb{C}$ *quasi-nearly subharmonic* if $u \in L^1_{loc}(\Omega)$ and if there is a constant $K = K(u, \Omega) \geqslant 1$ such that

$$u(z) \leqslant \frac{K}{r^2} \int_{B^2(z,r)} u(w) \, dm(w) \tag{6.3}$$

for each disk $B^2(z, r) \subset \Omega$.

In [135], the term *pseudoharmonic functions* is used, while in [122], the condition (6.3) is called sh_K-condition. If $K = 1$ and if u takes its values in $[-\infty, \infty]$, then u is called *nearly subharmonic* (see [66]).

Let $QNS_K(\Omega)$ denote the class of all functions satisfying (6.3) (for a fixed K), and by $QNS(\Omega)$ the class of all quasi-nearly subharmonic functions defined in Ω; so

$$QNS(\Omega) = \bigcup_{K \geqslant 1} QNS_K(\Omega).$$

The following fact generalizes the above-mentioned result of Fefferman and Stein [44] and gives one of the most important properties of QNS.

Theorem 6.1 *If $u \in QNS_K(\Omega)$ and $p > 0$, then $u^p \in QNS_C(\Omega)$, where C is a constant depending only on p and K. In particular, if u^p is quasi-nearly subharmonic for some $p > 0$, then it is so for every $p > 0$.*

The list of relevant papers related to Theorem 6.1 includes the following [3, 38, 122, 125, 134–136].

Theorem 6.2 ([41, Theorem 2.3] Koebe One-Quarter Theorem) *Let φ be a conformal mapping from the disk $B^2(z_0, R)$ into $\mathbb{C}$. Then the image $\varphi(B^2(z_0, R))$ contains the disk $B^2(\varphi(z_0), \rho)$, where $\rho = R|\varphi'(z_0)|/4$.*

Theorem 6.3 ([41, Theorem 2.5] Koebe Distortion Theorem) *Let φ be a conformal mapping from the disk $B^2(z_0, R)$ into $\mathbb{C}$. Then there holds inequalities*

$$\frac{R^2(R - |z - z_0|)}{(R + |z - z_0|)^3} \leqslant \frac{|\varphi'(z)|}{|\varphi'(z_0)|} \leqslant \frac{R^2(R + |z - z_0|)}{(R - |z - z_0|)^3}, \quad z \in B^2(z_0, R).$$

Consequently, if $|z - z_0| < R/2$, then

$$\frac{|\varphi'(z)|}{|\varphi'(z_0)|} \geqslant \frac{4}{27}.$$

Theorem 6.4 ([87, Theorem 1]) *Let $u \in QNS_K(\Omega)$. If φ is a conformal mapping from a domain G onto Ω, then the composition $u \circ \varphi$ belongs to the $QNS_C(G)$, where C depends only on K.*

Proof Let $u \in QNS_K(\Omega)$ and let φ be a conformal mapping from G onto Ω. We will find a constant C such that

$$\int_{B^2(z_0,r)} u(\varphi(z)) dm(z) \geqslant u(\varphi(z_0)) r^2 / C, \tag{6.4}$$

for $r < dist(z, \partial G)$. Let $w_0 = \varphi(z_0)$ and $\rho = r|\varphi'(z_0)|/4$, and let $\psi : \Omega \longrightarrow G$ denote the inverse of φ. Then

$$\int_{B^2(z_0,r)} u(\varphi(z)) dm(z) = \int_{\varphi(B^2(z_0,r))} u(w)|\psi'(w)|^2 dm(w)$$

$$\geqslant \int_{B^2(w_0,\rho/2)} u(w)|\psi'(w)|^2 dm(w),$$

where Theorem 6.2 above has been applied. Apply now Theorem 6.3 to the function ψ and get $|\psi'(w)| \geqslant (4/27)|\psi'(w_0)|$, for $|w - w_0| < \rho/2$. Hence

$$\int_{B^2(z_0,r)} u(\varphi(z)) dm(z) \geqslant (4/27)^2 |\psi'(w_0)|^2 \int_{B^2(w_0,\rho/2)} u(w) \, dm(w)$$

$$\geqslant (4/27)^2 |\psi'(w_0)|^2 (\rho/2)^2 u(w_0)/K$$

$$= (4/27)^2 |\psi'(w_0)|^2 |\varphi'(z_0)|^2 u(w_0) r^2 / (16K).$$

We now use the identity $\psi'(w_0)\varphi'(z_0) = 1$ and get (6.4) with $C = 27^2 K$. This concludes the proof. □

6.2 Regularly Oscillating Functions and Conformal Mappings

A function $f : \Omega \longrightarrow \mathbb{R}$ defined in $\Omega \subseteq \mathbb{C}$ is called *regularly oscillating* (see [125]) if f is of class $C^1(\Omega)$ and if for some K,

$$|\nabla f(z)| \leqslant Kr^{-1} \sup_{B^2(x,r)} |f - f(z)|, \quad B^2(z,r) \subseteq \Omega. \tag{6.5}$$

In [122] the class of such functions is denoted by $OC_K^1(\Omega)$ (O = oscillation). The class of all regularly oscillating functions will be denoted by $RO(\Omega)$.

Theorem 6.5 ([122, Theorem 3]) *If f is regularly oscillating, then $|f|$ and $|\nabla f|$ are quasi-nearly subharmonic. Moreover, if $f \in OC_K^1(\Omega)$, then $|f|$ and $|\nabla f|$ are in $QNS_C(\Omega)$, where C depends only on K.*

Example 6.6 Harmonic functions are regularly oscillating.

Example 6.7 ([122]) Convex functions are regularly oscillating. It follows that the modulus of the gradient of a convex function is quasi-nearly subharmonic.

We are ready now for the following theorem.

Theorem 6.8 ([87, Theorem 2]) *If $f \in RO(\Omega)$, and φ is a conformal mapping from G onto Ω, then $f \circ \varphi \in RO(G)$. Moreover, if $f \in OC_K^1(\Omega)$, then $f \circ \varphi$ is in $OC_{K_1}^1(\Omega)$, where K_1 depends only on K.*

Proof Let $u \in OC_K^1(\Omega)$ and let φ be a conformal mapping from G onto Ω. We will find a constant K_1 such that for $B^2(z_0, \varepsilon) \subset G$, we have

$$|\nabla u(\varphi(z_0))| \cdot |\varphi'(z_0)| \leqslant \frac{K_1}{\varepsilon} \sup_{z \in B^2(z_0, \varepsilon)} |u(\varphi(z)) - u(\varphi(z_0))|. \qquad (6.6)$$

Let ψ be the inverse of φ and let $w_0 = \varphi(z_0)$ and $\rho = \varepsilon |\varphi'(z_0)|/4$. Then recalling the definition of OC_K^1 and applying Theorem 6.2, we get

$$\begin{aligned}
&\sup\{|u(\varphi(z)) - u(\varphi(z_0))| \,:\, z \in B^2(z_0, \varepsilon)\} \\
&= \sup\{|u(w) - u(w_0)| \,:\, w \in \varphi(B^2(z_0, \varepsilon))\} \\
&\geqslant \sup\{|u(w) - u(w_0)| \,:\, w \in B^2(w_0, \rho)\} \\
&\geqslant |\nabla u(w_0)| \rho / K \\
&= |\nabla u(w_0)| \cdot |\varphi'(z_0)| \varepsilon / (4K).
\end{aligned}$$

This gives us (6.6) with $K_1 = 4K$. The proof is completed. □

6.3 Quasi-Nearly Subharmonic Functions and QC Mappings

If h is a function harmonic in a domain Ω in the Euclidean space $\mathbb{R}^n$, then the function $|h|^p$, which need not be subharmonic in Ω for $0 < p < 1$, behaves like a subharmonic function in the sense that the inequality

$$|h(a)|^p \leq \frac{C}{r^n} \int_{B^n(a,r)} |h|^p \, dm, \quad B^n(a, r) \subset \Omega, \, 0 < p < \infty, \qquad (6.7)$$

holds for some constant $C \geq 1$, where $B^n(a, r) = \{x : |x - a| < r\}$ and where m is the Lebesgue measure normalized so that $|\mathbb{B}^n| := m(\mathbb{B}^n) = 1$. The constant C in (6.7) depends only on n and p when $p < 1$, and $C = 1$ when $p \geq 1$.

From the Fefferman–Stein proof it follows that (6.7) remains true if $|h|$ is replaced by a nonnegative subharmonic function. So we have the following result.

Theorem 6.9 *If $u \geq 0$ is a function subharmonic in a domain $\Omega \subset \mathbb{R}^n$, then*

$$u(a)^p \leq \frac{C}{r^n} \int_{B^n(a,r)} u^p \, dm, \quad B^n(a,r) \subset \Omega, \ 0 < p < \infty, \tag{6.8}$$

where C depends only on p and n, when $p < 1$, and $C = 1$ when $p \geq 1$.

Remark 6.10 The list of references related to Theorem 6.9 includes the following: [3, 38, 122, 125, 134–136].

Definition 6.11 Let $u \geq 0$ be a locally bounded, measurable function on Ω. We say (see [122, 125]) that u is *C-quasi-nearly subharmonic* (abbreviated C-QNS) if the following condition is satisfied:

$$u(a) \leq \frac{C}{r^n} \int_{B^n(a,r)} u \, dm, \quad \text{whenever } B^n(a,r) \subset \Omega. \tag{6.9}$$

One can view (6.9) as a weak sub-mean value property. Besides nonnegative subharmonic functions, it also holds for nonnegative subsolutions to a large family of second order elliptic equations, see [64]. In fact, (6.9) is typically proven as a step towards Harnack inequalities for second order elliptic equations, using the Moser iteration scheme. Notice that u is quasi-nearly subharmonic if and only if u is everywhere dominated by its centered minimal function [36].

The paper [37] gives a partial generalization of the invariance under conformal mappings for both function classes QNS and RO, a result originally proven in [87]. More precisely, O. Dovgoshey and J. Riihentaus have shown in [37] that in $\mathbb{R}^n$, $n \geq 2$, both classes QNS and RO are invariant under bi-Lipschitz mappings. Since we know that if f is L-bi-Lipschitz, then it is L^{2n-2}-quasiconformal, it is natural to try to generalize this result to the whole class of quasiconformal mappings, and even more generally, to the class of all quasiregular mappings of bounded multiplicity.

To prove our next theorem we need the following lemmas.

Lemma 6.12 ([133, Proposition 4.14, pp. 20–21]) *Let $f : G \longrightarrow \mathbb{R}^n$ be quasiregular. Then the transformation formula*

$$\int_E (h \circ f) J_f \, dm = \int_{\mathbb{R}^n} h(y) N(y, f, E) \, dy$$

holds for every measurable $h : \mathbb{R}^n \longrightarrow [0, \infty]$ and all measurable $E \subseteq G$.

Lemma 6.13 *Let $f : \Omega \to \Omega'$ be K-quasiregular and of bounded multiplicity N. Let $x \in \Omega$ and $0 < r \leq \frac{1}{2}d(x, \partial\Omega))$. Then*

$$d\big(f(x), \partial f(B^n(x,r))\big) \geq \delta \sup_{y\in\overline{B}^n(x,r)} |f(y) - f(x)|,$$

where $\delta = \delta(n, K, N)$.

Proof Let $x \in \Omega$ and let $0 < r \leq \frac{1}{2}d(x, \partial\Omega)$. Note that $f(x)$ is an interior point of $f(B^n(x,r))$, because f is open. Moreover

$$\sup_{y\in\overline{B}^n(x,r)} |f(y) - f(x)| = |f(z) - f(x)|$$

for some $z \in \partial B^n(x,r)$, and

$$0 < d(f(x), \partial f(B^n(x,r)) = |f(\omega) - f(x)|$$

for some $\omega \in \partial B^n(x,r)$.

Let $E = [f(x), f(\omega)]$ be the segment between $f(w)$ and $f(\omega)$, and F be a segment that joins $f(z)$ to $\partial\Omega'$ (or to infinity) outside the ball

$$B^n(f(x), |f(z) - f(\omega)|).$$

We may assume that

$$|f(z) - f(x)| \geq 2|f(\omega) - f(x)|.$$

Let

$$u(y) = \begin{cases} 1, & y \in \overline{B}^n(f(x), |f(\omega) - f(x)|), \\ 0, & y \in (B^n(f(x), |f(z) - f(x)|))^c, \\ \log\frac{|f(z)-f(x)|}{|y-f(x)|} \Big/ \log\frac{|f(z)-f(x)|}{|f(\omega)-f(x)|} & \text{elsewhere.} \end{cases}$$

Then, by Lemma 6.12, the following holds:

$$\int_\Omega |\nabla(u\circ f)|^n\, dm \leq K \int_{\Omega'} |(\nabla u)(f(y))|^n J_f(y)\, dm(y)$$
$$\leq KN \int_{\Omega'} |\nabla u|^n\, dm$$
$$\leq \frac{KNC_n}{\log^{n-1}\frac{|f(z)-f(x)|}{|f(\omega)-f(x)|}}.$$

Note that since f is open, the set $f^{-1}([f(x), f(\omega)])$ contains a continuum joining x to $\partial B^n(x,r)$, and the set $f^{-1}(F)$ a contains a continuum joining $\partial B^n(x,r)$ to $\partial B^n(x, \frac{3}{2}r)$. By the capacity estimates (see, e.g., [89, p. 59]) we have

$$\int_\Omega |\nabla(u\circ f)|^n\, dm \geq \delta_0(n, K) > 0.$$

Now the lemma follows.   □

The following lemma is an immediate consequence of Lemma 6.13.

Lemma 6.14 *Let* $f : B^n(0, 2) \to \Omega'$ *be a K-quasiregular mapping with bounded multiplicity N such that* $f(0) = 0$. *Then there exist* $\rho \in (0, 1)$ *and* $R > 0$ *such that* $B^n(0, R) \supset f(\mathbb{B}^n) \supset B^n(0, \rho)$, *where* $R/\rho \le 1/\delta$, *and* δ *depends only on* $K, n,$ *and* N.

We shall also need the following fundamental fact.

Lemma 6.15 ([63, p. 258])   *Under the hypotheses of Lemma 6.14, there exists* $p > 1$ *such that*

$$\left(\int_{\mathbb{B}^n} J(y, f)^p \, dm \right)^{1/p} \le C \int_{\mathbb{B}^n} J(y, f) \, dm,$$

where p depends only on K, N, and n.

We now need to introduce the following norm.

Definition 6.16 (QNS-Norm)   Let

$$\|u\|_{\text{QNS}} = \inf \left\{ C \ge 0 : u(a) \le \frac{C}{r^n} \int_{B^n(a,r)} u \, dm \text{ for all } a \in \Omega', \, 0 < r \le d(a, \partial\Omega') \right\}.$$

Theorem 6.17 ([90, Theorem 1.2])   *If* $u \ge 0$ *is a C-quasi-nearly subharmonic function defined on a domain* $\Omega' \subset \mathbb{R}^n$, $n \ge 2$, *and* f *is a K-quasiregular mapping with bounded multiplicity N from a domain* Ω *onto* Ω', *then the function* $u \circ f$ *is* C_1-*quasi-nearly subharmonic in* Ω, *where* C_1 *only depends on* $K, C, N,$ *and* n.

Proof   The proof reduces to the case $\Omega = B^n(0, 2)$, $f(0) = 0$. Let $v = u \circ f$. Using rotations and translations, it can be seen that it is enough to prove that

$$v(0) \le C \int_{\mathbb{B}^n} v(y) \, dm(y),$$

for $C = C(K, n, \|u\|_{\text{QNS}})$. By Theorem 6.1, it suffices to find $q = q(K, N, n) \ge 1$ such that

$$v(0) \le C \left(\int_{\mathbb{B}^n} v(y)^q \, dm(y) \right)^{1/q}. \tag{6.10}$$

Towards the proof of this, we start from the Hölder inequality:

$$\int_{\mathbb{B}^n} v(y) J(y, f) \, dm(y) \le \left(\int_{\mathbb{B}^n} v(y)^q \, dm(y) \right)^{1/q} \left(\int_{\mathbb{B}^n} J(y, f)^p \, dm(y) \right)^{1/p},$$

where $p = q/(q-1)$. Using Lemma 6.12, we have

$$
\int_{\mathbb{B}^n} v(y) J(y, f) \, dm(y) = \int_{f(\mathbb{B}^n)} u(y) N(y, f, \mathbb{B}^n) \, dm(y)
$$

$$
\geq \int_{f(\mathbb{B}^n)} u(y) \, dm(y)
$$

$$
\geq \int_{B^n(0,\rho)} u(y) \, dm(y)
$$

$$
\geq c\rho^n u(0) = c\rho^n v(0).
$$

Note that we have used here Lemma 6.14 and the hypothesis that u is QNS. On the other side, by Lemmas 6.15 and 6.14, we have

$$
\left(\int_{\mathbb{B}} J(y, f)^p \, dm(y) \right)^{1/p} dy \leq C \int_{\mathbb{B}^n} J(y, f) \, dm(y)
$$

$$
= C \int_{f(\mathbb{B}^n)} N(y, f, \mathbb{B}^n) \, dm(y)
$$

$$
\leq CN |f(\mathbb{B}^n)|
$$

$$
\leq CN |B^n(0, R)| = CN k_n R^n.
$$

Combining these inequalities, we obtain

$$
c\rho^n v(0) \leq CN k_n R^n \left(\int_{\mathbb{B}^n} v(y)^q \, dm(y) \right)^{1/q}.
$$

Hence

$$
v(0) \leq \frac{CN k_n R^n}{c\rho^n} \left(\int_{\mathbb{B}^n} v(y)^q \, dm(y) \right)^{1/q}.
$$

The desired result now follows from the inequality $R/\rho \leq 1/\delta$, where δ depends only on $K, n,$ and N.   □

It should be noted that the hypothesis of bounded multiplicity of f in Theorem 6.17 is necessary as the following example shows. Let $f(z) = e^z$, $\Omega = \mathbb{C}$, $\Omega' = \mathbb{C} \setminus \{0\}$,

$$
E = \bigcup_{j \geq 2} [\exp(2^j), \exp(2^j + 1)],
$$

and $u(w) = \chi_E(|w|)$. Then it can be checked that u is quasi-nearly subharmonic in Ω' but, on the other hand, $u \circ f$ is not quasi-nearly subharmonic in Ω.

We shall now consider the above morphism property in more detail.

Definition 6.18 Let Ω and Ω' be subdomains of $\mathbb{R}^n$. A mapping $f : \Omega \to \Omega'$ is a *quasi-nearly subharmonic morphism* (QNS-morphism) if there is a constant $C < \infty$ such that for every quasi-nearly subharmonic u defined in Ω' we have

$$\|u \circ f\|_{\text{QNS}} \le C\|u\|_{\text{QNS}},$$

where $\|u\|_{\text{QNS}}$ is the quasi-norm defined above in Definition 6.16. If the inequality holds with a constant C, we call f a C-quasi-nearly subharmonic-morphism. Finally, f is a *strong QNS-morphism* if there is a constant C so that f restricted to each domain $G \subset \Omega$, $f : G \to G'$, is a C-QNS-morphism.

Theorem 6.19 ([90, Theorem 1.3]) *Let Ω, $\Omega' \subset \mathbb{R}^n$, $n \ge 2$, be domains. Then a homeomorphism $f : \Omega \to \Omega'$ is a strong QNS-morphism if and only if f is quasiconformal.*

Proof It is an immediate consequence of Theorem 6.17 that the quasiconformality of the homeomorphism f is a sufficient condition for f to be a strong QNS-morphism.

For the other direction, it suffices to prove that $f^{-1} : \Omega' \to \Omega$ is quasiconformal. Thus it suffices to verify the existence of $H < \infty$ such that

$$\limsup_{r \to 0} \frac{\text{diam}\,(f^{-1}(\overline{B}^n(y, r)))^n}{|f^{-1}(\overline{B}^n(y, r))|} \le H \tag{6.11}$$

for all $y \in \Omega'$ (see page 64 of [62]).

To simplify our notation, we write $x' = f(x)$ for $x \in \Omega$ in what follows.

Fix $y' \in \Omega'$ and let $r > 0$. Towards proving (6.11), we may assume that r is sufficiently small such that

$$B^n(y, 2\,\text{diam}\,(f^{-1}(\overline{B}^n(y', 2r)))) \subset \Omega.$$

Fix $y'_0 \in \partial B^n(y', 2r)$ and pick $y'_1 \in \partial B^n(y', r)$ so that

$$|y'_0 - y'_1| = \max_{\omega' \in \partial B^n(y', r)} |\omega' - y'_0|.$$

Set $G' = \Omega' \setminus \{y'_0\}$ and $G = \Omega \setminus \{y_0\}$. Now $B^n(y_1, |y_1 - y_0|/2) \subset G$ and

$$\text{diam}\,(f^{-1}(\overline{B}^n(y', r)) \le 2|y_1 - y_0|. \tag{6.12}$$

Define $u(\omega') = \chi_{\overline{B}^n(y', r)}(\omega')$ for $\omega' \in G'$. Then u is QNS in G' with $\|u\|_{\text{QNS}} \le 3^n$. Since f is C-QNS-morphism in G, we conclude that

$$u \circ f(y_1) \le C3^n \frac{1}{|B^n(y_1, |y_1 - y_0|/2)|} \int_{B^n(y_1, |y_1 - y_0|/2)} u \circ f \, dm$$
$$= C3^n \frac{|f^{-1}(\overline{B}^n(y'; r)) \cap B^n(y_1, |y_1 - y_0|/2)|}{|B^n(y_1, |y_1 - y_0|/2)|}$$

Recalling (6.12) and that $u \circ f(y_1) = u(y_1') = 1$, we arrive at

$$\mathrm{diam}\, (f^{-1}(\overline{B}^n(y', r))^n \leq CC_n |f^{-1}(\overline{B}^n(y', r))|$$

as desired.   □

It should be noted that if we assume the $W_{\mathrm{loc}}^{1,n}$ regularity for f, a version of Theorem 6.19 holds also for QNS-morphism.

The reader may wish to compare this with related quasiconformal invariance properties for other function classes: BMO-functions, weight functions of the class A_∞, and doubling measures; see [18, 131, 148, 149, 154].

To introduce A_∞-measures, we need the following theorem of Coifman and Fefferman:

Theorem 6.20 ([35]) *When Q is a cube with sides parallel to coordinates axis and μ is a measure defined on the Borel sets of $\mathbb{R}^n$, the following conditions are equivalent:*

1. *There exist $\delta_1 > 0$ and $C_1 > 0$ such that for all measurable $E \subseteq Q$ the following holds:*

$$\mu(E)/\mu(Q) \leqslant C_1(|E|/|Q|)^{\delta_1}.$$

2. *There exist $\delta_2 > 0$ and $C_2 > 0$ such that for all measurable $E \subseteq Q$ the following holds:*

$$|E|/|Q| \leqslant C_2(\mu(E)/\mu(Q))^{\delta_2}.$$

3. *$d\mu = w(x)\, dx$ and there exist $C > 0$ and $a > 0$ such that for all Q*

$$|Q|^{-1} \int_Q w(x)\, dx \leqslant C\left(|Q|^{-1} \int_Q w(x)^{-a}\, dx\right)^{-1/a}.$$

Definition 6.21 The class of all measures μ that satisfy the conditions of the previous theorem is denoted by A_∞.

In his paper [154] Uchiyama gave a characterization of quasiconformal mappings using A_∞ measures by proving that if a homeomorphism of $\mathbb{R}^n$ φ is ACL and differentiable a.e., then φ is quasiconformal if and only if for all $\mu \in A_\infty$, the measures $\mu \circ \varphi$ and $\mu \circ \varphi^{-1}$ are also in A_∞.

Now we are ready to introduce the BMO class of functions.

Definition 6.22 A locally integrable real valued function u is said to be of *bounded mean oscillation* (BMO) in $\mathbb{R}^n$, if

$$\frac{1}{|Q|} \int_Q \left| u(x) - \frac{1}{|Q|} \int_Q u(x)\, dx \right| dx \leqslant K$$

for every cube $Q \subseteq \mathbb{R}^n$ and some constant K.

On the space of BMO-functions, the BMO-norm can be defined by

$$||u||_{BMO} = \sup_{Q \subseteq \mathbb{R}^n} \frac{1}{|Q|} \int_Q |u(x) - u_Q| dx,$$

where

$$u_Q = \frac{1}{|Q|} \int_Q u(x) \, dx.$$

It should be noted that Uchiyama's characterization is based on the result of Reimann [131], where under the similar conditions on φ (with the additional assumption that $|\varphi(\cdot)|$ and $|\varphi^{-1}(\cdot)|$ are absolutely continuous set functions) it follows that φ is quasiconformal iff there exists $C > 0$ such that

$$||f \circ \varphi^{-1}||_{BMO} \leqslant C ||f||_{BMO} \tag{6.13}$$

for each BMO function f.

Later Astala proved that the local variant of (6.13) holds without analytic conditions on homeomorphism φ. More precisely, he proved the following theorem:

Theorem 6.23 ([18]) *Let* $\varphi : G \to G'$ *be an orientation preserving homeomorphism. If there exists a constant C such that*

$$\frac{1}{C} ||u||_{BMO,D'} \leqslant ||u \circ \varphi||_{BMO,D} \leqslant C ||u||_{BMO,D'} \tag{6.14}$$

holds for all subdomains $D \subseteq G$ and for all $u \in BMO(D')$, $D' = \varphi(D)$, then φ is quasiconformal.

Definition 6.24 Let D be a domain in $\mathbb{R}^n$ and let μ be a Borel measure defined on D. Let $2Q$ denote the cube concentric with Q and of the side length twice that of Q. We say that μ is a *doubling* on D and write $\mu \in \mathscr{D}(D)$ if there exists a constant $c > 0$ such that $\mu(2Q) \leqslant c \, \mu(Q)$ for all cubes Q with $2Q \subseteq D$.

In his paper [149], Staples gave a characterization of quasiconformal mappings $f : G \to G'$ in terms of doubling measures using, additionally to ACL and differentiability a.e. of f, that f and f^{-1} satisfy the condition (N) (that the image of a null set is a null set), and the assumption that $\nu = \mu \circ f$ is a doubling measure on $D \subseteq G$ for every doubling measure μ on $D' = f(D)$. Similarly as in Theorem 6.23 this is a local type condition that is assumed on every subdomain D of G.

Theorem 6.25 ([90, Theorem 1.4]) *Let $n \geq 2$ and let $f : \Omega \to \Omega'$ be a quasi-nearly subharmonic-morphism that belongs to $W_{loc}^{1,n}$. If, additionally, $J(x, f) \geq 0$ almost everywhere, then f is quasiregular.*

Proof Our definition of quasiregular mappings requires them to be continuous; however, this condition is superfluous here and it suffices to show that there exists $K \geq 1$ so that

$$|Df(x)|^n \leq KJ(x, f)$$

holds almost everywhere; see e.g., [133, p. 177]. Furthermore, every mapping f with Sobolev regularity $W_{loc}^{1,1}$ is approximatively differentiable almost everywhere. More precisely, for almost every x_0 and every $\epsilon > 0$, the set

$$A_\epsilon = \left\{ x : \frac{|f(x) - f(x_0) - Df(x_0)(x - x_0)|}{|x - x_0|} < \epsilon \right\}$$

has density one at x_0; see, e.g., [165, p. 140]. Since we are assuming Sobolev regularity $W_{loc}^{1,n}$, it suffices to show that the above distortion inequality holds at every such point x_0.

We only give the proof in the planar case, assuming differentiability instead of approximate differentiability. The higher dimensional case and the switch to approximate differentiability require technical modifications that should be evident to any reader who examine the argument below.

We start by assuming that f is differentiable at x_0.

Case (a) The function f is differentiable at x_0 with $J_f(x_0) \neq 0$. Then in an appropriate coordinate systems we have

$$Df(x_0) = \begin{bmatrix} a & 0 \\ 0 & b \end{bmatrix}.$$

Assume $0 < a < b$. We will show that b/a is bounded. Consider the function

$$u(\omega) = \chi_{\{\omega = x' + iy' : 0 \leq y' \leq x'\}}(\omega - f(x_0)).$$

If y_0 is in the interior of the sector $S = \{x' + iy' : 0 \leq y' \leq x'\}$ (i.e., $y_0 = 1 + i/2$), and $0 < r < d(y_0, \partial S)$, then $u(\omega) = 1$ holds for all $\omega \in B^n(f(x_0) + y_0, r)$. Also, $0 \leq u \leq 1$. Therefore, $\|u\|_{QNS} = 1$. So

$$T^{-1}\left(\{\omega = x' + iy' : 0 \leq y' \leq x'\}\right) = \{z = x + iy : 0 \leq by \leq ax\}$$
$$= \{z = x + iy : 0 \leq y \leq (a/b)x\}$$

for the linear transformation T associated with $Df(x_0)$. Thus $u \circ f(x_0) = 1$. If $r > 0$ is such that $B^n(x_0, r) \subset \Omega$, we conclude from the morphism property of f that

$$
\begin{aligned}
1 &\le \frac{C}{r^2} \int_{B^n(0,r)} u \circ f \, dm \\
&= \frac{C}{r^2} \frac{1}{2} r^2 \arctan \frac{a}{b} + o(r) \\
&\to \frac{C}{2} \frac{a}{b},
\end{aligned}
$$

when $r \to 0$, where $C > 0$ comes from the morphism property. Hence $b/a \le C/2$.

Case (b) Suppose now that f is differentiable at x_0 with $J_f(x_0) = 0$. We want to prove that $Df(x_0) = 0$. We argue by contradiction: suppose $Df(x_0) \ne 0$. We may assume

$$
Df(x_0) = \begin{bmatrix} 0 & 0 \\ 0 & 1 \end{bmatrix}.
$$

Define

$$
u(\omega) = \chi_{\{\omega = x' + iy' : 0 \le |y'| \le x'\}}(\omega - f(x_0)).
$$

Then $\|u\|_{\text{QNS}} = 1$, and

$$
f^{-1}(t + it + f(x_0)) = s_1(t) + i s_2(t) + x_0,
$$

where $\lim\limits_{t \to 0} \dfrac{s_2(t)}{s_1(t)} = 0$. But then there is no $C > 0$ such that

$$
(u \circ f)(x_0) \le \frac{C}{r^2} \int_{B^n(x_0,r)} u \circ f \, dm
$$

for all small $r > 0$, which contradicts the morphism property of f.

$\square$

Each quasiregular mapping f is either constant or both open and discrete; in the latter case the multiplicity of f is locally finite. The condition $J(x, f) \ge 0$ in Theorem 6.25 cannot be dropped, as the mapping $f(x, y) = (x, |y|)$ is a planar (strong) quasi-nearly subharmonic-morphism. It turns out that the Sobolev regularity assumption can be slightly relaxed: if f above is a C-quasi-nearly subharmonic-morphism, then local p-integrability of the distributional derivatives suffices for $p = p(n, C) < n$. This can be deduced using [69] and the proof of Theorem 6.25. In the injective planar case even the regularity $W_{\text{loc}}^{1,1}$ suffices.

6.4 Regularly Oscillating Functions and QC Mappings

In this section we emphasize the invariance of a related function class, introduced in [122].

Definition 6.26 A function $u : \Omega' \to \mathbb{R}^k$ is said to be *regularly oscillating* if

$$\operatorname{Lip} u(x) \le Cr^{-1} \sup_{y \in B^n(x,r) \subset \Omega'} |u(y) - u(x)|, \quad x \in \Omega', \ B^n(x,r) \subset \Omega', \quad (6.15)$$

where $C \ge 0$ is a constant independent of x and r. Here

$$\operatorname{Lip} u(x) = \limsup_{y \to x} \frac{|u(y) - u(x)|}{|y - x|}.$$

(We borrow this notation from [89].) Note that $\operatorname{Lip} u(x) = |\operatorname{grad} u(x)|$ if u is differentiable at x. The smallest C satisfying (6.15) will be denoted by $\|u\|_{RO}$.

We will now show the following invariance property of regularly oscillating functions.

Theorem 6.27 ([90, Theorem 1.5]) *Let $f : \Omega \to \Omega'$ be quasiregular, regularly oscillating, and of bounded multiplicity in Ω. If u is regularly oscillating in Ω', then $u \circ f$ is regularly oscillating in Ω with $\|u \circ f\|_{RO} \le C' \|u\|_{RO}$, where C' depends only on the multiplicity of f, K, n, and $\|f\|_{RO}$.*

Proof Let $x \in \Omega$ and $0 < r < \frac{1}{2} d(x, \partial \Omega)$. Since the mapping f is regularly oscillating and quasiregular we have, by Lemma 6.13,

$$\operatorname{Lip} f(x) \le Cr^{-1} \sup_{y \in \overline{B}^n(x,r)} |f(y) - f(x)|$$
$$\le \frac{C}{\delta} r^{-1} d(f(x), \partial f(B^n(x,r))).$$

We now recall that nonconstant quasiregular mappings are open. Since u is regularly oscillating and $d(f(x), \partial f(B(x,r))) > 0$, we have that

$$\operatorname{Lip} u(f(x)) \le \hat{C} d(f(x), \partial f(B^n(x,r)))^{-1} \sup_{z \in B^n(f(x), d(f(x), \partial f(B^n(x,r))))} |u(f(x)) - z|$$
$$\le \hat{C} d(f(x), \partial f(B^n(x,r)))^{-1} \sup_{y \in B^n(x,r)} |u \circ f(y) - u \circ f(x)|.$$

It follows that

$$\operatorname{Lip}(u \circ f)(x) \le \operatorname{Lip}(u(f(x)) \operatorname{Lip} f(x)$$
$$\le \hat{C} d(f(x), \partial f(B^n(x,r)))^{-1} \sup_{y \in B^n(x,r)} |u \circ f(y) - u \circ f(x)|$$
$$\times \frac{C}{\delta} r^{-1} d(f(x), \partial f(B^n(x,r)))$$
$$= C' r^{-1} \sup_{y \in B^n(x,r)} |u \circ f(y) - u \circ f(x)|.$$

This completes the proof of the theorem. $\qquad\qquad\qquad\qquad\qquad\qquad\square$

It should be noted that the assumption that f is regularly oscillating is necessary in this theorem. This can be seen by observing that the coordinate projections are regularly oscillating. Recall that not all quasiregular mappings are regularly oscillating. However, in the case of analytic functions, this assumption can be dropped. Moreover, similarly as in Theorem 6.25, quasiregularity is also necessary if we assume that $J(x, f) \geq 0$ almost everywhere. However, no Sobolev regularity is needed because regularly oscillating functions and mappings are locally Lipschitz continuous.

In case $n = 2$ and when f is conformal the invariance property of Theorem 6.27 was established in [87].

It should be also remarked that the assumption of bounded multiplicity of f in Theorem 6.27 is necessary as in the case of Theorem 6.17. To see this simply let $f(z) = e^z$, $\Omega = \mathbb{C}$, $\Omega' = \mathbb{C} \setminus \{0\}$,

$$E = \bigcup_{j \geq 2}[\exp(2^j), \exp(2^j + 1)],$$

and $v(w) = \int_0^{|w|} \chi_E(t)\, dt$. Then v is regularly oscillating but $v \circ f$ is not.

6.5 Some Generalizations and Examples

It is natural to pose the following question:

Question 6.28 Can these results be generalized for metric spaces?

Recently, in the paper [38], Dovgoshey and Riihentaus gave some generalizations of Definition 6.11 as well as some examples related to these generalizations.

Let Ω be a domain in $\mathbb{R}^n$, $n \geq 2$.

Definition 6.29 ([38], see also the reference therein) Let $u : \Omega \to [-\infty, \infty)$ be a Lebesgue measurable function on Ω. We say that u is *C-quasi-nearly subharmonic in narrow sense* for a constant $C \geq 1$ if $u^+ \in \mathscr{L}^1_{loc}(\Omega)$ and the following condition is satisfied:

$$u(a) \leq \frac{C}{|B^n(a,r)|}\int_{B^n(a,r)} u\, dm, \quad \text{whenever } \overline{B^n(a,r)} \subseteq \Omega. \tag{6.16}$$

A function u is *quasi-nearly subharmonic in narrow sense* if it is C-quasi-nearly subharmonic in narrow sense for at least one $C \geq 1$.

Definition 6.30 ([38]) Let $u : \Omega \to [-\infty, \infty)$ be a Lebesgue measurable function on Ω. We say that u is *C-quasi-nearly subharmonic* if $u^+ \in \mathscr{L}^1_{loc}(\Omega)$ and for all $M \geq 0$ the following condition is satisfied:

$$u_M(a) \leq \frac{C}{|B^n(a,r)|} \int_{B^n(a,r)} u_M\, dm, \quad \text{whenever } B^n(a,r) \subset \Omega, \tag{6.17}$$

where $u_M(x) = \max\{u(x), -M\} + M$. A function u is *quasi-nearly subharmonic* if it is C-quasi-nearly subharmonic for at least one $C \geq 1$.

The class of functions from Definition 6.11 is a proper subclass of the class of functions from Definition 6.29. The class of functions from Definition 6.29 is a proper subclass of the class of functions from Definition 6.30. However, all three definitions match in the class of nonnegative functions.

Example 6.31 ([38]) The function $u : \mathbb{R}^2 \to \mathbb{R}$

$$u(x,y) = \begin{cases} -1, & y < 0, \\ 1, & y \geq 0 \end{cases}$$

is 2-quasi-nearly subharmonic, but not quasi-nearly subharmonic in a narrow sense.

Example 6.32 ([38]) The function $u : \mathbb{R}^2 \to \mathbb{R}$

$$u(x,y) = \begin{cases} 3, & x = 0, \\ 1, & x \neq 0 \end{cases}$$

is 3-quasi-nearly subharmonic. The constant function $v : \mathbb{R}^2 \to \mathbb{R}$, $v(x,y) = -2$ is harmonic. Then

$$(u+v)(x,y) = \begin{cases} 1, & x = 0, \\ -1, & x \neq 0 \end{cases}$$

and

$$(u+v)_M = \max\{u+v, -M\} + M = (u+v+M)^+$$

for every $M \geq 0$. In particular for $M = 1$ we obtain

$$(u+v)_1(x,y) = \begin{cases} 2, & x = 0, \\ 0, & x \neq 0. \end{cases}$$

Since $(u+v)_1(0,0) > 1$ and the double integral $\iint_{B^2(a,r)} (u+v)_1(x,y)\, dx\, dy$ is zero for every $a \in \mathbb{R}^2$ and $r > 0$, the function $(u+v)_1$ is not quasi-nearly subharmonic. Hence $(u+v)$ is also not quasi-nearly subharmonic.

Chapter 7
Possible Research Directions

7.1 Characterizations of Boundary Values

We have already mentioned that Pavlović [123] made a deep and detailed analysis of the boundary values of harmonic quasiconformal mappings of the unit disk $\mathbb{D}$ by proving the following theorem.

Theorem 7.1 *Let u be a harmonic homeomorphism of $\mathbb{D}$ that extends continuously to the boundary, and $f : S^1 \to S^1$ be its boundary function. Then the following conditions are equivalent:*

1. *u is quasiconformal;*
2. *u is bi-Lipschitz in the Euclidean metric;*
3. *f is bi-Lipschitz and the Hilbert transformation of its derivative is in L^∞.*

This result has initiated an extensive line of research between the theories of bi-Lipschitz conditions and HQC mappings (see [15, 79, 118] and [76]).

In [99], the author has shown that HQC mappings between *every* two proper domains in the plane are bi-Lipschitz with respect to the corresponding quasihyperbolic metrics. More precisely, in [99] the following result was obtained.

Theorem 7.2 *Suppose D and D' are proper domains in $\mathbb{R}^2$. If $f : D \to D'$ is K-quasiconformal and harmonic, then it is bi-Lipschitz with respect to the quasihyperbolic metrics on D and D'.*

The proof of this result was based on a counterpart of Koebe theorem, established by Astala and Gehring, using estimates of a geometric notion of average derivative on a ball

$$\alpha_f(z) = \exp\frac{1}{n}(\log J_f)_{B_z}. \tag{7.1}$$

© Springer Nature Switzerland AG 2019

V. Todorčević, *Harmonic Quasiconformal Mappings and Hyperbolic Type Metrics*,
https://doi.org/10.1007/978-3-030-22591-9_7

This leads us to the following natural problem about the co-Lipschitz condition in higher dimensions.

Problem 7.3 Does a HQC map have a nonvanishing Jacobian in dimensions greater than two?

It should be noted that this problem is closely related to properties of the function (7.1), so this will be the first thing to be examined here.

Another line of research initiated by Theorem 7.1 deals with characterizations of boundary values of HQC mappings. In [79], the analogous result was proved for the half-plane. A natural question arises: what happens in higher dimensions, both in the case of the unit ball and in the case of the upper half space? Since the answer to this question in dimension $n = 2$ involves Hilbert's transform, we expect that in higher dimensions it will involve singular integral operators, like Riesz transforms.

A variant of this problem is to replace the condition of harmonic quasiconformal (HQC) with the condition of harmonic quasiregular (HQR), as well as with the condition of p-harmonic quasiconformal mappings, which involve nonlinear potential theory. Namely, p-harmonic mappings minimize the L^p norm of the gradient similar to the situation of L^2 norm under standard harmonic mappings. This provides possibilities to extend our research to quasiconformal aspects of potential analysis, a very active area of current research.

The characterization of boundary values can also be approached in terms of intrinsic metric properties of the domain. In particular, hyperbolic and quasihyperbolic metrics, which depend on the geometric properties of the domain in $\mathbb{R}^n$, capture some of the most essential properties relevant for boundary behavior. A map $f : X \rightarrow Y$ can then be analyzed as a map between corresponding metric spaces (X, d_X) and (Y, d_Y). We say that f is a (C, D) quasi-isometry if

$$\frac{1}{C} d_Y(f(x), f(y)) - D \leq d_X(x, y) \leq C d_Y(f(x), f(y)) + D.$$

A hyperbolic space $\mathbb{H}^n$ can be modeled as a unit ball $\mathbb{B}^n$, equipped with a hyperbolic metric. It has a boundary S^{n-1} at infinity. Each quasi-isometry $f : \mathbb{H}^3 \rightarrow \mathbb{H}^3$ extends continuously to S^2, and the restriction of this extension to S^2 is quasiconformal.

The question about existence of harmonic quasiconformal extension of a quasiconformal (quasisymmetric for dimension $n = 2$) boundary condition was a long-standing open problem. The famous Schoen conjecture (formulated in the Introduction) was proved by V. Marković in [108]. The analogue of this conjecture in dimension 3 is also proved by V. Marković [107] while analogues for dimensions $n > 3$ are proved by L. Marius and V. Marković [106].

Theorem 7.4 (Generalized Schoen Conjecture [106–108]) *Suppose that $f : S^{n-1} \rightarrow S^{n-1}$ is a quasiconformal map. Then there exists a unique harmonic (with respect to the hyperbolic metric) and quasi-isometric map $\hat{f} : \mathbb{H}^n \rightarrow \mathbb{H}^n$ that extends f.*

The two-dimensional case is important because of the links with the Teichmüller theory. Because of the uniqueness part of this conjecture, proved by P. Li and L. F. Tam in [96], harmonic quasiconformal hyperbolic mappings can serve as representatives of a class in the Universal Teichmüller space.

Using quasi-isometries, which are a natural and more flexible generalization of quasiconformal mappings, it seems possible that there is another possible approach leading us to the following question:

Problem 7.5 Characterize boundary mappings $f : S^{n-1} \to S^{n-1}$ for which the Euclidean harmonic extension $\hat{f} : \mathbb{B}^n \to \mathbb{B}^n$ is quasi-isometric with respect to the hyperbolic metric.

For general domains, the quasihyperbolic metric is comparable to the usual hyperbolic metric in a simply connected plane domain by the Koebe distortion theorem, and the quasihyperbolic metric continues to be a useful tool in dimensions $n \geq 3$, whereas the hyperbolic metric is less useful. Also, it seems natural to consider quasi-isometries with respect to quasihyperbolic metric as a condition to replace quasiconformality. Thus, we propose a more general question in higher dimensions:

Problem 7.6 Characterize boundary mappings $f : \partial D \to \partial D'$ for which there exists a unique Euclidean harmonic map $\hat{f} : D \to D'$, quasi-isometric with respect to the quasihyperbolic metrics, that extends f, where D and D' are proper domains in $\mathbb{R}^n$.

7.2 QC and HQC Mappings on Non-smooth Domains

The geometry of domains has been an important topic in Geometric Function Theory. Martio and Sarvas [112] were the first to consider and introduce the so-called *uniform domains* in the late 1970s. We recall that these are domains in which every two points can be joined by an arc of length bounded by a constant times the Euclidean distance between them such that each point on that arc divides the arc into two parts, at least one of them of length smaller than a constant times the distance of dividing point to the boundary of the domain (i.e., the arc does not come too close to the boundary of the domain). Around the same time, Jones [72] characterized domains which have the property that bounded mean oscillation (BMO) functions can be extended from the domain to the whole space. Gehring and Osgood in [53] have proved that Jones domains are precisely uniform domains.

Uniform domains equipped with the quasihyperbolic metric have the property of being Gromov hyperbolic. In the eighties the Russian mathematician M. Gromov introduced the notion of hyperbolic space, which has thereafter been studied and developed further by many authors. For a long time the research was centered at Hyperbolic Group Theory, but lately many researchers in Geometric Function Theory have developed interests towards the theory of Gromov hyperbolic spaces.

In particular, Bonk, Heinonen, and Koskela [29] have established a canonical correspondence between geodesic Gromov hyperbolic spaces and non-complete uniform
domains. The non-geodesic case has been considered by Bonk and Schramm [28]
and by Väisälä [156].

The main result of the author's paper [91] with P. Koskela and P. Lammi provides
characterization of Gromov hyperbolicity of the quasihyperbolic metric spaces
(Ω, k) by geometric properties of the Ahlfors regular length metric measure space
(Ω, d, μ). The characterizing properties are called the Gehring–Hayman condition
and ball-separation condition.

The investigation of an Ahlfors regular domain in the context of the HQC
mappings can be deepened. An *Ahlfors regular domain* D is a domain $D \subset \mathbb{C}$
which is bounded by an *Ahlfors regular curve*, , i.e., a rectifiable curve γ such that
any small disk of radius r covers part of γ of length at most Cr for some $C > 0$.

Consider mappings $f : \mathbb{D} \longrightarrow \Omega$, where $\mathbb{D} = \{z \in \mathbb{C} : |z| < 1\}$ and Ω is a
simply connected domain in the complex plane.

Theorem 7.7 *Suppose* $f : \mathbb{D} \to \Omega$ *is conformal and* Ω *is an arbitrary simply
connected domain. Then*

$$\log f'(z) \in \mathscr{B},$$

where $\mathscr{B}$ *denotes the Bloch space, consisting of analytic functions* $g : \mathbb{D} \to \mathbb{C}$ *with*

$$\|g\|_{\mathscr{B}} := \sup_{|z|<1} (1 - |z|^2)|g'(z)| < \infty.$$

We note that in this context it is known that $\mathscr{B} \equiv \{g \in BMO(\mathbb{D}) : g \text{ analytic}\}$
(see, e.g., [127]).

Suppose that Ω is a *chord-arc domain*. This means that $J := \partial\Omega$ is a chord-arc
curve, i.e., J is a rectifiable Jordan curve and there is a constant $C < \infty$ such that

$$l(a, b) \leq C|a - b| \qquad \forall a, b \in J.$$

Here $l(a, b)$ denotes the length of the shorter of the arcs of J connecting a and b.
Chord-arc domains are also called *Lavrentiev domains*, and they are precisely the
images of a circle or a line under a bi-Lipschitz mapping of $\mathbb{C}$. Then by a result of
Pommerenke (see, e.g., his book [127, pp. 155]), we have the following:

Theorem 7.8 *Suppose* $f : \mathbb{D} \to \Omega$ *is conformal with* Ω *a chord-arc domain. Then*

$$\log f'(z) \in BMOA. \tag{7.2}$$

Here $BMOA := \{f|_{\partial\mathbb{D}} \in BMO(\partial\mathbb{D}) : f \text{ analytic in } \mathbb{D}\}.$

Many further results in this direction have been established. In particular, (7.2) is shown by proving that $|f'| \in A_p$ for some p; here A_p is the Muckenhoupt-class on the boundary $\partial \mathbb{D}$.

By Kellog's theorem (see, e.g., [47, p. 62]), we have

Theorem 7.9 *Suppose* $f : \mathbb{D} \to \Omega$ *is conformal, with* $\partial \Omega$ *a* $C^{1,\alpha}$-*Jordan curve. Then* $f \in C^1(\overline{\mathbb{D}})$ *with* $f'(z) \neq 0$ *in* $\overline{\mathbb{D}}$. *In particular,*

$$\log f'(z) \in \mathbb{H}^\infty,$$

so f *is bi-Lipschitz.*

This result is also true in the more general domains.

The next natural step of our interest here is a generalization to quasiconformal mappings. By a theorem of Reimann (see, e.g., [22, p. 347]), we have the following.

Theorem 7.10 *Suppose* $f : \mathbb{D} \to \Omega$ *is quasiconformal. Then*

$$\log J(z, f) \in BMO(\mathbb{D}).$$

Furthermore, if f is conformal, then $2\text{Re} \log f'(z) \equiv 2\log|f'(z)| \equiv \log J(z, f)$. Hence Reimann's result is the natural counterpart to Theorem 7.7. Some further analogies of $\log f'(z)$ and $\log J(z, f)$ have been studied in [20]. However, the Theorems 7.8 and 7.9 do not seem to have natural counterparts in the case of general quasiconformal mappings, since quasiconformal mappings $f : \mathbb{D} \to \mathbb{D}$ can be singular on the boundary $\partial \mathbb{D}$. This suggests that in the case of harmonic quasiconformal mappings we can combine the above results with the theorem of the author [99], which shows that $\log J(z, f)$ is superharmonic for every harmonic homeomorphism f. Hence from many respects, for harmonic quasiconformal mappings, the quantity $\log J(z, f)$ seems the natural counterpart to $\log f'(z)$. In particular, the question arises if the counterparts of Theorems 7.7, 7.8, and 7.9 hold for HQC mappings and $\log J(z, f)$. It should be noted that the first one holds and follows from Theorem 7.10 above.

There are many studies of bi-Lipschitzity of mappings, but properties of HQC mappings from a disk to an Ahlfors domain seem to be little studied. In fact M. Zinsmeister generalized Theorem 7.8 to domains with Ahlfors regular boundary [166]. From the other side D. Kalaj considered a generalization of Theorem 7.8 to harmonic quasiconformal mappings and proved several equivalent characterizations of the Muckenhoupt property of the derivative on the boundary [77]. Bounded chord-arc domains are exactly the bi-Lipschitz images of the unit disk. However the natural class of domains are the more general domains having an Ahlfors regular boundary. In fact, Ω is a chord-arc domain if and only if it is both a quasidisk and has Ahlfors regular boundary.

One possible goal for future research is to give a natural full generalization of the Zinsmeister theorem for HQC mappings.

Problem 7.11 If $f : \mathbb{D} \rightarrow \Omega$ is harmonic and quasiconformal and if Ω is an Ahlfors regular domain, is f absolutely continuous on the boundary?

A positive answer to the following problem would extend the result of D. Kalaj mentioned above.

Problem 7.12 Is the corresponding boundary measure induced by the mapping in the Muckenhoupt class A_p for some p?

References[1]

1. A. Abaob, M. Arsenović, M. Mateljević, A. Shkheam, Moduli of continuity of harmonic quasiconformal mappings on bounded domains. Ann. Acad. Sci. Fenn. Math. **38**(2), 839–847 (2013)
2. S.B. Agard, F.W. Gehring, Angles and quasiconformal mappings. Proc. Lond. Math. Soc. **s3-14a**, 1–21 (1965)
3. P. Ahern, J. Bruna, Maximal and area integral characterization of Hardy-Sobolev spaces in the unit ball of $\mathbb{C}^n$. Rev. Mat. Iberoamericana **4**(1), 123–153 (1988)
4. L. Ahlfors, *Möbius Transformations in Several Dimensions*. Contemporary Mathematics: Introductory Courses (Mir, Moscow, 1986), 112 pp.
5. L. Ahlfors, *Lectures on Quasiconformal Mappings*. University Lecture Series, vol. 38, 2nd edn. (American Mathematical Society, Providence, 2006)
6. L. Ahlfors, A. Beurling, Conformal invariants and function-theoretic null-sets. Acta Math. **83**, 101–129 (1950)
7. H. Aikawa, Hölder continuity of the Dirichlet solution for a general domain. Bull. Lond. Math. Soc. **34**(6), 691–702 (2002)
8. G.D. Anderson, Dependence on dimension of a constant related to the Grötzsch ring. Proc. Am. Math. Soc. **61**, 77–80 (1976)
9. G.D. Anderson, M.K. Vamanamurty, M. Vuorinen, Sharp distortion theorems for quasiconformal mappings. Trans. Am. Math. Soc. **305**(1), 95–111 (1988)
10. G.D. Anderson, M.K. Vamanamurthy, M.K. Vuorinen, *Conformal Invariants, Inequalities, and Quasiconformal Maps*. Canadian Mathematical Society Series of Monographs and Advanced Texts (Wiley, New York, 1997)
11. G.D. Anderson, M. Vuorinen, X. Zhang, *Topics in Special Functions III*. Analytic Number Theory, Approximation Theory and Special Functions (Springer, New York, 2014)
12. M. Arsenović, V. Manojlović, On the modulus of continuity of harmonic quasiregular mappings on the unit ball in R^n. Filomat **23**(3), 199–202 (2009)
13. M. Arsenović, V. Kojić, M. Mateljević, On Lipschitz continuity of harmonic quasiregular maps on the unit ball in $\mathbb{R}^n$. Ann. Acad. Sci. Fenn. Math. **33**(1), 315–318 (2008)
14. M. Arsenović, V. Manojlović, M. Mateljević, Lipschitz-type spaces and harmonic mappings in the space. Ann. Acad. Sci. Fenn. Math. **35**(2), 379–387 (2010)

---

[1] **Remark to the Reference List:** The author has published papers under surnames Kojić, Manojlović and Todorčević.

15. M. Arsenović, V. Božin, V. Manojlović, Moduli of continuity of harmonic quasiregular mappings in $\mathbb{B}^n$. Potential Anal. **34**(3), 283–291 (2011)
16. M. Arsenović, V. Manojlović, M. Vuorinen, Hölder continuity of harmonic quasiconformal mappings. J. Inequal. Appl. **2011**, 37 (2011)
17. M. Arsenović, V. Manojlović, R. Näkki, Boundary modulus of continuity and quasiconformal mappings. Ann. Acad. Sci. Fenn. Math. **37**(1), 107–118 (2012)
18. K. Astala, A remark on quasiconformal mappings and BMO-functions. Mich. Math. J. **30**(2), 209–212 (1983)
19. K. Astala, F.W. Gehring, Quasiconformal analogues of theorems of Koebe and Hardy-Littlewood. Mich. Math. J. **32**(1), 99–107 (1985)
20. K. Astala, F.W. Gehring, Injectivity, the BMO norm and the universal Teichmüller space. J. Anal. Math. **46**, 16–57 (1986)
21. K. Astala, V. Manojlović, On Pavlović's theorem in space. Potential Anal. **43**(3), 361–370 (2015)
22. K. Astala, T. Iwaniec, G.J. Martin, *Elliptic Partial Differential Equations and Quasiconformal Mappings in the Plane* (Princeton University Press, Princeton, 2009)
23. S. Axler, P. Bourdon, W. Ramey, *Harmonic Function Theory*. Graduate Texts in Mathematics, vol. 137 (Springer, New York, 1992)
24. A.F. Beardon, *The Geometry of Discrete Groups*. Graduate Texts in Mathematics, vol. 91 (Springer, New York, 1995)
25. A.F. Beardon, The Apollonian metric of a domain in $\mathbb{R}^n$, in *Quasiconformal Mappings and Analysis*, ed. by P. Duren, J. Heinonen, B. Osgood, B. Palka (Springer, New York, 1998), pp. 91–108
26. H.P. Boas, Julius and Julia: mastering the art of the Schwarz lemma. Am. Math. Mon. **117**(9), 770–785 (2010)
27. B. Bonfert-Taylor, R. Canary, E.C. Taylor, Quasiconformal homogeneity after Gehring and Palka. Comput. Methods Funct. Theory **14**(2), 417–430 (2014)
28. M. Bonk, O. Schramm, Embeddings of Gromov hyperbolic spaces. Geom. Funct. Anal. **10**(2), 266–306 (2000)
29. M. Bonk, J. Heinonen, P. Koskela, *Uniformizing Gromov Hyperbolic Spaces*. Astérisque, vol. 270 (American Mathematical Society, Providence, 2001)
30. B. Branner, N. Fagella, *Quasiconformal Surgery in Holomorphic Dynamics*. Cambridge Studies in Advanced Mathematics, vol. 141 (Cambridge University Press, Cambridge, 2014)
31. J. Byström, Sharp constants for some inequalities connected to the p-Laplace operator. J. Inequal. Pure. Appl. Math. **6**(2), 56A, 8 pp. (2005)
32. L.A. Caffarelli, D. Kinderlehrer, Potential methods in variational inequalities. J. Anal. Math. **37**, 285–295 (1980)
33. P. Caraman, *n-Dimensional Quasiconformal (QCf) Mappings* (Abacus Press, Tunbridge Wells, 1974)
34. L. Carleson, *Selected Problems on Exceptional Sets*. Van Nostrand Mathematical Studies, vol. 13 (Van Nostrand, Princeton, 1967)
35. R.R. Coifman, C. Fefferman, Weighted norm inequalities for maximal functions and singular integrals. Stud. Math. **51**, 241–250 (1974)
36. D. Cruz-Uribe, The minimal operator and the geometric maximal operator in $\mathbb{R}^n$. Stud. Math. **144**(1), 1–37 (2001)
37. O. Dovgoshey, J. Riihentaus, Bi-Lipschitz mappings and quasinearly subharmonic functions. Int. J. Math. Math. Sci. **2010**, 382179, 8 pp. (2010)
38. O. Dovgoshey, J. Riihentaus, On quasinearly subharmonic functions. Lobachevskii J. Math. **38**(2), 245–254 (2017)
39. O. Dovgoshey, P. Hariri, M. Vuorinen, Comparison theorems for hyperbolic type metrics. Complex Var. Elliptic Equ. **61**(11), 1464–1480 (2016)
40. P. Duren, *Theory of H^p Spaces*. Pure and Applied Mathematics, vol. 38 (Academic, New York, 1970)

41. P. Duren, *Univalent Functions*. Fundamental Principles of Mathematical Sciences, vol. 259 (Springer, New York, 1983)
42. P. Duren, *Harmonic Mappings in the Plane* (Cambridge University Press, Cambridge, 2004)
43. R. Estrada, M. Pavlović, L'hospital's monotone rule, Gromov's theorem and operations that preserve the monotonicity of quotients. Publ. Inst. Math. **101**(115), 11–24 (2017)
44. C. Fefferman, E.M. Stein, H^p-spaces of several variables. Acta Math. **129**(3–4), 137–193 (1972)
45. B. Fuglede, Extremal length and functional completion. Acta Math. **98**, 171–219 (1957)
46. J.B. Garnett, *Bounded Analytic Functions* (Academic, New York, 1981)
47. J.B. Garnett, D.E. Marshall, *Harmonic Measure* (Cambridge University Press, Cambridge, 2005)
48. F.W. Gehring, Symmetrization of rings in space. Trans. Am. Math. Soc. **101**, 499–519 (1961)
49. F.W. Gehring, Rings and quasiconformal mappings in space. Trans. Am. Math. Soc. **103**, 353–393 (1962)
50. F.W. Gehring, The L^p-integrability of the partial derivatives of a quasiconformal mapping. Acta Math. **130**, 265–277 (1973)
51. F.W. Gehring, Quasiconformal mappings in Euclidean spaces, in *Handbook of Complex Analysis: Geometric Function Theory*, vol. 2 (Elsevier, Amsterdam, 2005), pp. 1–29
52. F.W. Gehring, K. Hag, The Ubiquitous Quasidisk. Mathematical Surveys and Monographs, vol. 184 (AMS, Providence, 2010)
53. F.W. Gehring, B.G. Osgood, Uniform domains and the quasihyperbolic metric. J. Anal. Math. **36**, 50–74 (1979)
54. F.W. Gehring, B.P. Palka, Quasiconformally homogeneous domains. J. Anal. Math. **30**, 172–199 (1976)
55. D. Gilbarg, N. Trudinger, *Elliptic Partial Differential Equations of Second Order* (Springer, Berlin, 1998)
56. S. Gleason, T. Wolff, Lewy's harmonic gradient maps in higher dimensions. Commun. Partial Differ. Equ. **16**(12), 1925–1968 (1991)
57. G.H. Hardy, J.E. Littlewood, Some properties of conjugate functions. J. Reine Angew. Math. **167**, 405–423 (1933)
58. P. Hariri, M. Vuorinen, X. Zhang, Inequalities and bi-Lipschitz conditions for the triangular ratio metric. Rocky Mountain J. Math. **47**(4), 1121–1148 (2017)
59. P. Hästö, A new weighted metric: the relative metric. I. J. Math. Anal. Appl. **274**, 38–58 (2002)
60. P. Hästö, The Apollonian metric and biLipschicity mappings, Thesis (Ph.D.) – Helsingin Yliopisto (2003), 49 pp.
61. P. Hästö, Z. Ibragimov, D. Minda, S. Ponnusamy, S.K. Sahoo, Isometries of some hyperbolic-type path metrics and the hyperbolic medial axis. In the tradition of Ahlfors-Bers, IV. Contemp. Math. **432**, 63–74 (2007)
62. J. Heinonen, P. Koskela, Definitions of quasiconformality. Invent. Math. **120**(1), 61–79 (1995)
63. J. Heinonen, P. Koskela, Weighted Sobolev and Poincaré inequalities and quasiregular mappings of polynomial type. Math. Scand. **77**, 251–271 (1995)
64. J. Heinonen, T. Kilpeläinen, O. Martio, *Nonlinear Potential Theory of Degenerate Elliptic Equations*. Oxford Mathematical Monographs (Oxford University Press, New York, 1993)
65. S. Hencl, P. Koskela, *Lectures on Mappings of Finite Distortion*. Lecture Notes in Mathematics, vol. 2096 (Springer, Berlin, 2013)
66. M. Hervé, *Analytic and Plurisubharmonic Functions in Finite and Infinite Dimensional Spaces*. Lecture Notes in Mathematics, vol. 198 (Springer, Berlin, 1971)
67. J. Hesse, A p-extremal length and p-capacity equality. Ark. Mat. **13**, 131–141 (1975)
68. A. Hinkkanen, Modulus of continuity of harmonic functions. J. Anal. Math. **51**, 1–29 (1988)
69. T. Iwaniec, p-harmonic tensors and quasiregular mappings. Ann. Math. **136**(3), 589–624 (1992)
70. T. Iwaniec, J. Onninen, Radó–Kneser–Choquet theorem. Bull. Lond. Math. Soc. **46**(6), 1283–1291 (2014)

71. P. Järvi, M. Vuorinen, Uniformly perfect sets and quasiregular mappings. J. Lond. Math. Soc. **54**(3), 515–529 (1996)

72. P.W. Jones, Extension domains for BMO. Indiana Univ. Math. J. **29**, 41–66 (1980)

73. D. Kalaj, Quasiconformal harmonic functions between convex domains. Publ. Inst. Math. **75**(89), 139–146 (2004)

74. D. Kalaj, Quasiconformal and harmonic mappings between Jordan domains. Math. Z. **260**(2), 237–252 (2008)

75. D. Kalaj, Lipschitz spaces and harmonic mappings. Ann. Acad. Sci. Fenn. Math. **34**(2), 475–485 (2009)

76. D. Kalaj, A priori estimate of gradient of a solution to certain differential inequality and quasiconformal mappings. J. Anal. Math. **119**, 63–88 (2013)

77. D. Kalaj, Muckenhoupt weights and Lindelöf theorem for harmonic mappings. Adv. Math. **280**, 301–321 (2015)

78. D. Kalaj, V. Manojlović, Subharmonicity of the modulus of quasiregular harmonic mappings. J. Math. Anal. Appl. **379**(2), 783–787 (2011)

79. D. Kalaj, M. Pavlović, Boundary correspondence under quasiconformal harmonic diffeomorphisms of a half-plane. Ann. Acad. Sci. Fenn. Math. **30**(1), 159–165 (2005)

80. L. Keen, N. Lakić, *Hyperbolic Geometry from a Local Viewpoint*. London Mathematical Society, Student Texts, vol. 68 (Cambridge University Press, Cambridge, 2007)

81. R. Klén, On hyperbolic type metrics, Dissertation, University of Turku, Helsinki. Annales Academiae Scientiarum Fennicae Mathematica Dissertationes No. 152 (2009)

82. R. Klén, H. Lindén, M. Vuorinen, G. Wang, The visual angle metric and Möbius transformations. Comput. Methods Funct. Theory **14**(2–3), 577–608 (2014)

83. R. Klén, M. Vuorinen, X. Zhang, Quasihyperbolic metric and Möbius transformations. Proc. Am. Math. Soc. **142**(1), 311–322 (2014)

84. R. Klén, V. Todorčević, M. Vuorinen, Teichmüller's problem in space. J. Math. Anal. Appl. **455**(2), 1297–1316 (2017)

85. M. Knezević, M. Mateljević, On the quasi-isometries of harmonic quasiconformal mappings. J. Math. Anal. Appl. **334**, 404–413 (2007)

86. V. Kojić, Metric spaces and quasiconformal mappings. Master Thesis, Belgrade (2007)

87. V. Kojić, Quasi-nearly subharmonic functions and conformal mappings. Filomat **21**(2), 243–249 (2007)

88. V. Kojić, M. Pavlović, Subharmonicity of $|f|^p$ for quasiregular harmonic functions, with applications. J. Math. Anal. Appl. **342**(1), 742–746 (2008)

89. P. Koskela, *Lectures on Quasiconformal and Quasisymmetric Mappings* (University of Jyväskylä, Jyväskylä, 2009)

90. P. Koskela, V. Manojlović, Quasi-nearly subharmonic functions and quasiconformal mappings. Potential Anal. **37**(2), 187–196 (2012)

91. P. Koskela, P. Lammi, V. Manojlović, Gromov hyperbolicity and quasihyperbolic geodesics. Ann. Sci. Éc. Norm. Supér. **47**(5), 975–990 (2014)

92. R. Kühnau, ed., *Handbook of Complex Analysis: Geometric Function Theory*, vol. 1/2 (Elsevier, Amsterdam, 2002/2005)

93. E.C. Lawrence, *Partial Differential Equations*. Graduated Studies in Mathematics, vol. 19 (American Mathematical Society, Providence, 1998)

94. O. Lehto, K.I. Virtanen, *Quasiconformal Mappings in the Plane*. Die Grundlehren der math. Wissenschaften, vol. 126, 2nd edn. (Springer, Berlin, 1973)

95. H. Lewy, On the non-vanishing of the Jacobian of a homeomorphism by harmonic gradients. Ann. Math. **88**, 518–529 (1968)

96. P. Li, L.F. Tam, Uniqueness and regularity of proper harmonic maps. Ann. Math. **137**(1), 167–201 (1993)

97. H. Lindén, Quasihyperbolic geodesics and uniformity in elementary domains. Dissertation, University of Helsinki, Helsinki, Annales Academiae Scientiarum Fennicae Mathematica Dissertationes No. 146 (2005), 50 pp.

98. V. Manojlović, Moduli of continuity of quasiregular mappings. Ph.D. Thesis, Belgrade (2008)

99. V. Manojlović, Bi-Lipschicity of quasiconformal harmonic mappings in the plane. Filomat **23**(1), 85–89 (2009)

100. V. Manojlović, BiLipschitz mappings between sectors in planes and quasi-conformality. Funct. Anal. Approx. Comput. **1**(2), 1–6 (2009)

101. V. Manojlović, On conformally invariant extremal problems. Appl. Anal. Discrete Math. **3**(1), 97–119 (2009)

102. V. Manojlović, Harmonic quasiconformal mappings in domains in $\mathbb{R}^n$. J. Anal. **18**(1), 297–316 (2010)

103. V. Manojlović, On biLipschicity of quasiconformal mappings. Novi Sad J. Math. **45**(1), 105–109 (2015)

104. V. Manojlović, V. Vuorinen, On quasiconformal maps with identity boundary values. Trans. Am. Math. Soc. **363**(5), 2467–2479 (2011)

105. A. Marden, S. Rickman, Holomorphic mapping of bounded distortion. Proc. Am. Math. Soc. **46**, 226–228 (1984)

106. L. Marius, V. Marković, Heat flows on hyperbolic spaces. J. Differ. Geom. **108**(3), 495–529 (2018)

107. V. Marković, Harmonic maps between 3-dimensional hyperbolic spaces. Invent. Math. **199**(3), 921–951 (2015)

108. V. Marković, Harmonic maps and Schoen conjecture. J. Am. Math. Soc. **30**(3), 799–817 (2017)

109. G.J. Martin, B.G. Osgood, The quasihyperbolic metric and the associated estimates on the hyperbolic metric. J. Anal. Math. **47**, 37–53 (1986)

110. O. Martio, On harmonic quasiconformal mappings. Ann. Acad. Sci. Fenn. Ser. A I **425**, 10 pp. (1968)

111. O. Martio, R. Näkki, Boundary Hölder continuity and quasiconformal mappings. J. Lond. Math. Soc. **44**(2), 339–350 (1991)

112. O. Martio, J. Sarvas, Injectivity theorems in plane and space. Ann. Acad. Sci. Fenn. Math. Ser. A I **4**(2), 383–401 (1979)

113. O. Martio, S. Rickman, J. Väisälä, Definitions of quasiregular mappings. Ann. Acad. Sci. Fenn. Ser. A I **448**, 40 pp. (1969)

114. O. Martio, J. Väisälä, Quasihyperbolic geodesics in convex domains. II. Pure Appl. Math. Q. **7**(2), 395–409 (2011). Special Issue: In Honor of Frederick W. Gehring, Part II

115. O. Martio, S. Rickman, J. Väisälä, Distortion and singularities of quasiregular mappings. Ann. Acad. Sci. Fenn. A I **465**, 1–13 (1970)

116. M. Mateljević, Distortion of harmonic functions and harmonic quasiconformal quasi-isometry. Rev. Roumaine Math. Pures Appl. **51**(5–6), 711–722 (2006)

117. M. Mateljević, *Topics in Conformal, Quasiconformal and Harmonic Maps* (Zavod za udžbenike, Beograd, 2012)

118. M. Mateljević, M. Vuorinen, On harmonic quasiconformal quasi-isometries. J. Inequal. Appl. **2010**, 178732, 19 pp. (2010)

119. G.D. Mostow, Quasi-conformal mappings in n-space and the rigidity of the hyperbolic space forms. Publ. Math. Inst. Hautes Etudes Sci. **34**, 53–104 (1968)

120. R. Näkki, B. Palka, Asymptotic values and Hölder continuity of quasiconformal mappings. J. Anal. Math. **48**, 167–178 (1987)

121. D. Partyka, K. Sakan, On bi-Lipschitz type inequalities for quasiconformal harmonic mappings. Ann. Acad. Sci. Fenn. Math. **32**(2), 579–594 (2007)

122. M. Pavlović, On subharmonic behaviour and oscillation of functions on balls in $\mathbb{R}^n$. Publ. Inst. Math. **55**, 18–22 (1994)

123. M. Pavlović, Boundary correspondence under harmonic quasiconformal homeomorphisms of the unit disk. Ann. Acad. Sci. Fenn. Math. **27**(2), 365–372 (2002)

124. M. Pavlović, *Function Classes on the Unit Disc – An Introduction*. Studies in Mathematics, vol. 52 (De Gruyter, Berlin 2014)

125. M. Pavlović, J. Riihentaus, Classes of quasi-nearly subharmonic functions. Potential Anal. **29**(1), 89–104 (2008)

126. E.A. Poletsky, Holomorphic quasiregular mappings. Proc. Am. Math. Soc. **95**(2) , 235–241 (1985)
127. Ch. Pommerenke, *Boundary Behavior of Conformal Maps*. Fundamental Principles of Mathematical Sciences, vol. 299 (Springer, Berlin, 1992)
128. S. Ponnusamy, T. Sugawa, M.K. Vuorinen, *Proceedings of International Workshop on Quasiconformal Mappings and their Applications*, December 27, 2005–Jan 1, 2006, IIT Madras (2007)
129. T. Rado, P.V. Reichelderfer, *Continuous transformations in analysis. With an introduction to algebraic topology*. Die Grundlehren der mathematischen Wissenschaften in Einzeldarstellungen mit besonderer Berücksichtigung der Anwendungsgebiete, Bd. LXXV (Springer, Berlin, 1955), vii + 442 pp.
130. K. Rajala, The local homeomorphism property of spatial quasiregular mappings with distortion close to one. Geom. Funct. Anal. **15**(5), 1100–1127 (2005)
131. H.M. Reimann, Functions of bounded mean oscillation and quasiconformal mappings. Comment. Math. Helv. **49**, 260–276 (1974)
132. Yu.G. Reshetnyak, *Spatial Mappings with Bounded Distortion*. Izdat. "Nauka" Sibirsk (Otdelenie, Novosibirsk, 1982, in Russian)
133. S. Rickman, *Quasiregular Mappings*. Ergebnisse der Mathematik und ihrer Grenzgebiete (3) [Results in Mathematics and Related Areas (3)], vol. 26 (Springer, Berlin, 1993)
134. J. Riihentaus, On theorem of Avanssian-Arsove. Expo. Math. **7**, 69–72 (1989)
135. J. Riihentaus, Subharmonic functions: non-tangential and tangential boundary behavior, in *Function Spaces, Differential Operators and Non-linear Analysis* (Academy of Sciences Czech Republic Institute of Mathematics, Prague, 2000), pp. 229–238
136. J. Riihentaus, A generalized mean value inequality for subharmonic functions. Expo. Math. **19**(2), 187–190 (2001)
137. L.A. Rubel, A.L. Shields, B.A. Taylor, *Mergelyan Sets and the Modulus of Continuity*. Approximation Theory (Academic, New York, 1973), pp. 457–460
138. W. Rudin, *Real and Complex Analysis. Second Edition*. McGraw-Hill Series in Higher Mathematics (McGraw-Hill, New York, 1974). xii + 452 pp.
139. R. Schoen, The role of harmonic mappings in rigidity and deformation problems, in *Complex Geometry*. Lecture Notes in Pure and Applied Mathematics, vol. 143 (Osaka, 1990)
140. P. Seittenranta, Linear dilatation of quasiconformal maps in space. Duke Math. J. **91**(1), 1–16 (1998)
141. P. Seittenranta, Möbius-invariant metrics. Math. Proc. Cambridge Philos. Soc. **125**(3), 511–533 (1999)
142. S. Simić, Lipschitz continuity of the distance ratio metric on the unit disk. Filomat **27**(8), 1505–1509 (2013)
143. S. Simić, Some sharp Lipschitz constants for the distance ratio metric. J. Anal. **21**, 147–155 (2013)
144. S. Simić, Distance ratio metric on a half-plane. J. Math. Sci. Adv. Appl. **27**(1), 43–48 (2014)
145. S. Simić, Distance ratio metric on the unit disk. J. Adv. Math. **6**(3), 1056–1060 (2014)
146. S. Simić, M. Vuorinen, Lipschitz conditions and the distance ratio metric. Filomat **29**(9), 2137–2146 (2015)
147. S. Simić, M. Vuorinen, G. Wang, Sharp Lipschitz constants for the distance ratio metric. Math. Scand. **116**(1), 86–103 (2015)
148. S.G. Staples, Maximal functions, A_∞ measures and quasiconformal maps. Proc. Am. Math. Soc. **113**(3), 689–700 (1991)
149. S.G. Staples, Doubling measures and quasiconformal maps. Comment. Math. Helv. **67**, 119–128 (1992)
150. E. Stein, *Singular Integrals and Differentiability Properties of Functions* (Princeton University Press, Princeton, 1970)
151. P.M. Tamrazov, Contour and solid structural properties of holomorphic functions of a complex variable. Uspehi Mat. Nauk **28**, 131–161 (1973, in Russian). English translation in Russian Math. Surveys 28 (1973), 141–173

152. O. Teichmüller, A displacement theorem of quasiconformal mapping, in *Handbook of Teichmüller Theory*. Translated from the German by Manfred Karbe. IRMA Lectures in Mathematics and Theoretical Physics, vol. VI (European Mathematical Society, Zürich, 2016), p. 27

153. V. Todorčević, Subharmonic behavior and quasiconformal mappings. Anal. Math. Phys. https://doi.org/10.1007/s13324-019-00308-8

154. A. Uchiyama, Weight functions of the class (A_∞) and quasi-conformal mappings. Proc. Jpn. Acad. **51**, 811–814 (1975)

155. J. Väisälä, *Lectures on n-Dimensional Quasiconformal Mappings*. Lecture Notes in Mathematics, vol. 229 (Springer, Berlin, 1971)

156. J. Väisälä, Gromov hyperbolic spaces. Expo. Math. **23**(3), 187–231 (2005)

157. M. Vuorinen, Conformal invariants and quasiregular mappings. J. Anal. Math. **45**, 69–115 (1985)

158. M. Vuorinen, *Conformal Geometry and Quasiregular Mappings*. Lecture Notes in Mathematics, vol. 1319 (Springer, Berlin, 1988)

159. M. Vuorinen, Quadruples and spatial quasiconformal mappings. Math. Z. **205**(4), 617–628 (1990)

160. M. Vuorinen, X. Zhang, Distortion of quasiconformal mappings with identity boundary values. J. Lond. Math. Soc. **90**(3), 637–653 (2014)

161. M. Wolf, The Teichmüller theory of harmonic maps. J. Differ. Geom. **29**(2), 449–479 (1989)

162. J.C. Wood, Lewy's theorem fails in higher dimensions. Math. Scand. **69**(2), 166 (1991)

163. X. Zhang, Hyperbolic type metrics and distortion of quasiconformal mappings, Ph.D. Thesis, University of Turku 2013

164. W.P. Ziemer, Extremal length and p-capacity. Mich. Math. J. **16**, 43–51 (1969)

165. W.P. Ziemer, *Weakly Differentiable Functions. Sobolev Spaces and Functions of Bounded Variation*. Graduate Texts in Mathematics, vol. 120 (Springer, New York, 1989)

166. M. Zinsmeister, Conformal representation and almost Lipschitz curves.Ann. Inst. Fourier **34**(2), 29–44 (1984)

Index

A

Absolute (cross) ratio, 64
Absolutely continuous mapping, xvi
Absolutely continuous on lines, 15
Admissible metric, 6
Ahlfors regular curve, 150
Ahlfors regular domain, 150
A_∞-measures, 140
A_∞-Muckenhoupt class, 126
Angle distortion, 76
Anticonformal mapping, xiv
Appolonian metric, 73
Arc length element, 63
Average derivative, 111
A_∞-weight, 126
A_1-weight, 126
A_p-weight, 126

B

Basic ellipticity bounds, 125
Bernoulli inequality, xvii, 72
Bi-Lipschitz property, 112
Bloch norm bounds, 125
Bloch space, 150
BMO-functions, 141
BMO-norm, 141
Boundary Harnack inequality, 115
Bounded mean oscillation, 140

C

Capacity, 47
C-bi-Lipschitz, xiv

Chordal metric, 57, 60
Chord-arc curve, 150
Chord-arc domain, 150
C-Lipschitz, xiv
Co-Lipschitz property, 120
Condenser, 47
Condition (N), 141
Conformal homeomorphism, xiv
Conformal invariants, 73
C-quasiisometry, xiv
C-quasi-nearly subharmonic, 145
C-quasi-nearly subharmonic in a narrow sense, 145
C-quasi-nearly subharmonic-morphism, 139
Cylinder example, 9

D

Degenerated ring example, 11
Diffeomorphism, xv
Distance ratio metric j_D, 67
Distortion theorem, 26
Doubling measure, 141

E

Eigenvalue, 52
Euclidean length of a curve, 70
Extremal property of Teichmüller ring, 28

F

Flat boundary, 118

© Springer Nature Switzerland AG 2019
V. Todorčević, *Harmonic Quasiconformal Mappings and Hyperbolic Type Metrics*,
https://doi.org/10.1007/978-3-030-22591-9

Printed in the United States
By Bookmasters